United States
Department of
Agriculture

Forest Service

Research and
Development

Gen. Tech.
Report
WO-79

May 2010

USDA

A Dynamic Invasive Species Research Vision: Opportunities and Priorities 2009–29

Edited by
Mary Ellen Dix and Kerry Britton

Acknowledgments

We thank Carlos Rodriguez-Franco (Forest Service) and Jim Reaves (Forest Service) for their guidance and support of the Invasive Species Strategic Program Area review process and development of the visionary papers. We acknowledge the authors for their contributions and dedication to the process.

We give special thanks to the peer reviewers of the papers for their critical observations and comments. We also thank Katrina Krause (Forest Service) and Marilyn Buford (Forest Service) for their sage advice.

Contents

Introduction

Invasive species significantly impact U.S. ecosystems and are one of the greatest threats to forest, rangeland, and urban forest health. They have contributed to increases in fire frequency and intensity; reduced water resources, forest growth, and timber; and negatively affected native species and their habitats throughout the United States. Global trade, climate change, and innovations in human transportation are just a few of the factors that have increased the rate of invasive species introduction and the costs associated with their prevention, quarantine, and management. Forest and rangeland managers urgently need effective management techniques to reduce invasive species' effects.

In 2006, the U.S. Department of Agriculture (USDA), Forest Service Research and Development Invasive Species Strategic Program Area (SPA) solicited programmatic feedback through a formal external peer review. The SPA used this information to guide long-term national planning for our research program. The National Research and Development Invasive Species Strategy was revised to reflect this guidance, and research efforts were prioritized to address the challenges managers will face in the future. Scientists developed 12 visionary papers that responded to our customers' feedback. These visionary papers identify future invasive species research issues and priorities and provide the Forest Service and its partners with a framework for programming and budgeting for the next 20 years.

Invasive Species Overarching Priorities to 2029

Kerry O. Britton[1,2], Marilyn Buford[3], Kelly Burnett[4], Mary Ellen Dix[5], Susan J. Frankel[6], Melody Keena[7], Mee-Sook Kim[8], Ned B. Klopfenstein[9], Michael E. Ostry[10], and Carolyn Hull Sieg[11]

Executive Summary

Invasive species are one of the greatest threats to forest, range, aquatic, and urban forest ecosystem health. They contribute to the endangerment of native species and may lead to other severe ecological and financial consequences in our Nation's wildlands and urban forests. Costs the public pays for damage, losses, and control efforts are estimated at more than $138 billion per year. Severe infestations of cheatgrass have contributed to increased fire frequency and intensity in Western States, reducing property values in some areas by up to 80 percent. Asian long-horned beetles threaten more than $500 billion in urban tree losses in America, over time, if left unchecked. Recent regional invasions, such as Sudden Oak Death in California, Emerald Ash Borer in the Midwest, and *Sirex noctilio* in New York, have the potential to become national threats. Invasive species threaten Pacific Island ecosystems, riparian communities, and wetlands and are the second leading contributor of species endangerment in aquatic ecosystems. Expanding global trade is increasing the rate of invasive species introductions and the costs associated with preventing introductions and quarantining and managing new infestations.

Given the large number of nonnative invasive species present in the United States and projections for increasing numbers, U.S. Department of Agriculture (USDA), Forest Service Research and Development (R&D) must be strategic in allocating research resources to develop the science to manage invasive species and their effects. Forest owners and managers likewise need tools to help allocate resources across pests and ecosystems. A recent peer review panel recommended increased funding for two areas: (1) prevention and prediction and (2) early detection and rapid response. Therefore, quantitative risk analysis and pathway assessments will be key components of our research program. Our future strategy also recognizes the importance of maintaining research in two other areas: (1) control and management and (2) restoration and rehabilitation. We believe a holistic national strategy will improve sharing of expertise across research stations and encourage actions that prevent regional threats from expanding into national ones. Increased coordination with other agencies will help identify regulatory and research gaps and will improve the complementary use of resources.

In response to the external peer review, Forest Service R&D identified research priorities for invasive species. Many of these priorities are described in the 12 papers that follow. Several overarching priorities, which are considered important for all invasive species, regardless of taxa, are discussed here.

1. **Quantify Invasive Species Biology, Ecology, Interactions, and Effects**

   a. Quantify genetic, ecological, and evolutionary relationships among high-priority invasive species and ecosystems where they occur.

---

[1] Authors listed in alphabetical order.

[2] National Forest Pathology Program Leader, Forest Service, Research and Development, 1601 North Kent St., Arlington, VA 22209.

[3] National Program Leader for Silviculture Research, Forest Service, Research and Development, 1601 North Kent St., Arlington, VA 22209.

[4] Research Fisheries Biologist, Forest Service, Pacific Northwest Research Station, 3200 SW Jefferson Way, Corvallis, OR 97331.

[5] National Invasive Species Research Program Lead, Forest Service, Research and Development, 1601 North Kent St., Arlington, VA 22209.

[6] Sudden Oak Death Program Manager, Forest Service, Pacific Southwest Research Station, 800 Buchanon St., Albany, CA 94710.

[7] Research Entomologist, Forest Service, Northern Research Station, 51 Mill Pond Rd., Hamden, CT, 06514.

[8] Assistant Professor, Kookmin University, Department of Forest Resources, 861-1 Jeongneung-Dong Seongbuk-Gu, Seoul, Korea 136-702.

[9] Research Plant Pathologist, Forest Service, Rocky Mountain Research Station, 1221 S. Main St., Moscow, ID 83843.

[10] Research Plant Pathologist, Forest Service, Northern Research Station, 1561 Lindig Ave., St. Paul, MN 55108.

[11] Research Plant Ecologist, Forest Service, Rocky Mountain Research Station, 2500 South Pine Knoll Dr., Flagstaff, AZ 86001.

b. Quantify ecological, social, and economic effects of invasive species.

c. Develop science-based protocols to prioritize invasive species to help managers assess action thresholds.

2. **Predicting and Prioritizing Invasive Species**

a. Develop a methodology for predicting which species are likely to become invasive.

b. Develop science-based protocols to prioritize prevention activities.

3. **Identifying and Detecting Invasive Species**

a. Improve invasive species detection and diagnostics technology.

b. Quantify invasive patterns and processes across geographical and elevational gradients.

4. **Managing Invasive Species and Altered Systems**

a. Develop more effective treatments and control or management methods for high-priority species.

b. Predict interactions between multiple invasive species and multiple disturbances under varying climatic scenarios.

This paper discusses the overarching research needs, anticipated products, outcomes, and skills needed to address these research priorities as part of the Forest Service R&D strategy for invasive species. The accompanying visionary papers identify priorities specific to individual taxa or issues closely linked with invasive species, such as prevention, disturbance, and economics.

Introduction

Forest Service Research and Development (R&D) has developed a more holistic view of our invasive species research program as we implement the National Strategy and Implementation Plan for Invasive Species (USDA Forest Service 2003). This agencywide strategy, which tiers from the National Invasive Species Council's Management Plan (National Invasive Species Council 2008) organizes our invasive species work in four broad activity areas: (1) prediction and prevention, (2) early detection and rapid response (EDRR), (3) control and management, and (4) rehabilitation

and restoration. Forest Service R&D develops and provides scientific information and tools to help accomplish these activities.

The scope of the Forest Service R&D invasive species research program is defined by research being conducted to predict, prevent, identify, detect, and mitigate the ecosystem effects of invasive species. Invasive species are defined in Executive Order 13112: "*Invasive species are those species that are not native to the ecosystem under consideration and whose introduction causes or is likely to cause economic or environmental harm. Invasive species include: plants, animals, fish, insects, diseases, invertebrates, and others.*" The Forest Service R&D invasive species research program includes research on species that are not native to the United States and those that are native but are advancing to invade other areas due to the increased connectivity of ecosystems and changing environmental conditions. Our research is conducted throughout the continental United States; in the tropical forests of Hawaii, the western Pacific, and Puerto Rico; and internationally. We conduct research at a variety of scales in wilderness, watersheds, old-growth forests, wetlands, aquatic systems, urban interface forests, grasslands, plantations, and utility corridors.

External Peer Review Panel Comments

In October 2006, a panel of external experts reviewed our invasive species research program. A summary of the panel's comments and suggestions in each of the four activity areas follows.

Prevention

The USDA Animal and Plant Health Inspection Service (APHIS) has the primary responsibility for preventing the introduction of invasive species into the United States. Forest Service R&D supports APHIS by developing and providing the scientific information and tools needed by Federal and State agencies for risk analysis, detection, monitoring, and interception so that APHIS can minimize or eliminate the potential for the introduction of invasive species. In fiscal year (FY) 2005, approximately 34 percent of our invasive species program nationwide focused on prevention. The external peer review panel recommended a proactive approach to prediction and prevention research by increasing interagency efforts especially in quantitative risk analysis, including pathway analysis, epidemiology, and socio-economic analysis. Because of long-term economic and environmental benefits, we anticipate an increase in prevention and prediction research as well as improved coordination among agencies to reduce overlap, identify regulatory and research gaps, and improve the complementary use of resources.

Early Detection and Rapid Response

Forest Service R&D supports the EDRR activities of APHIS and management agencies by providing scientific information and tools needed for detecting, monitoring, enforcing, and eradicating invasive species to prevent economic loss and ecological damage to the Nation's forests and rangelands. In FY 2005, approximately 14 percent of our invasive species research program nationwide focused on this activity. The external peer review panel recommended proactively expanding our efforts in EDRR and also expanding interagency coordination.

Control and Management

Forest Service R&D currently focuses approximately 36 percent of its invasive species research resources on reducing the extent and spread of established invasive species to minimize economic loss and ecological damage to the Nation's forests, rangelands, and watersheds. This activity area includes developing or evaluating management and mitigation treatment guidelines, tools, and systems; assessing tools for long-term efficacy and secondary effects; quantifying indirect and direct effects of invasive species and control mechanisms on ecosystems; and developing risk and cost-benefit analysis systems and prioritization tools. The external peer review panel suggested that we consider research on biological, chemical, and mechanical control systems according to the severity of environmental and economic risks and effects. Forest Service R&D mitigation and control research is expected to remain constant as a result of increasingly effective collaboration with external partners who are well situated to address particular elements of this research.

Restoration and Rehabilitation

Forest Service R&D focuses approximately16 percent of invasive species resources on developing tools for restoring, rehabilitating, and sustaining forest and rangeland ecosystems by preventing reinvasion and for regaining long-term multiple uses and values. Recognizing the disciplinary and human power needs to conduct this research, the panel recommended that we consider increasing our capacity by enhancing collaboration with external partners. Assessing interactions amongst multiple economic and ecological impacts and disturbances at varying spatial and temporal scales is crucial to developing tools to guide land management decisions about where to spend scarce resources.

Overarching Research Priorities

To provide responsive and critical science and technology in these activity areas, this paper outlines four overarching research priorities that Forest Service researchers developed: (1) quantifying invasive species biology, ecology, interactions, and effects; (2) predicting and prioritizing invasive species; (3) identifying and detecting invasive species; and (4) managing invasive species and altered systems. The research program guided by these priorities is relevant now, in the near future, and into the distant future and will produce key knowledge, tools, and management options.

Quantifying Invasive Species Biology, Ecology, Interactions, and Effects

Invasive species behavior depends on the genetic, ecological, and evolutionary relationships among interacting biological components and the physical environment. Quantifying these interactions forms the basis for developing prioritization tools, detection and control methods, and management options. The research must be conducted by highly interdisciplinary teams that include researchers from multiple research stations and external partners. Although site-level and short-term interactions among species will be of interest, the research will also address interactions over broader spatial and temporal extents. Primary areas of research include the following:

* Quantify the genetic, ecological, and evolutional relationships among high-priority invasive species and

ecosystems where they occur and determine their biology and behavior.

- Quantify ecological, social, and economic effects of invasive species.

- Develop science-based protocols to prioritize invasive species to help managers assess action thresholds.

Anticipated Products

Over the next 20 years, we anticipate concentrated efforts to support management goals that will result in the following:

- Basic understanding of the biology and behavior of specific invasive species necessary for developing prioritization, identification, and detection tools and control and management options.

- A comprehensive database of potentially invasive species with associated distribution data and identified geographic areas (ecosystems) at risk.

- Expanded databases containing DNA sequence information from diverse organisms and international locations to allow for more thorough comparisons among biological organisms worldwide.

- Comprehensive impact assessments, including ecological, economic, and societal values that help prioritize invasive species for research, management, and restoration activities.

Outcomes

Expected outcomes have been divided into a rough timeline, with the following expectations:

Now

Molecular tools and analytical methods are available to examine genetic and evolutionary relationships among ecosystem components, and these tools continue to be improved.

Sophisticated animal survey methods are now available and improving exponentially for determining vital multiple population parameters, such as species diversity and spatial variation. The latter two factors are critical to determining population size, which is paramount in managing for sustainable populations and maintaining biodiversity.

Some databases, such as the National Center for Biotechnology Information GenBank, climate models, and soil maps, are widely available and widely used in invasive species research.

Soon

We will maintain multidisciplinary taxonomic expertise and research on biological, genetic, and ecological attributes of species.

We will efficiently quantify genetic, ecological, and evolutionary relationships among the biological components of ecosystems through formalized broad-focus, collaborative efforts.

We will be able to project likely consequences of management actions on invasive species in a dynamic landscape.

We will be able to predict economic and ecological effects of key invasive species to help managers decide where to allocate resources.

Future

The Forest Service has a proud history of past excellence, and must strive to maintain our expert organization for long-term and global implications of invasive species in forest, rangeland, and associated aquatic ecosystems.

Integrated databases, advances in species and community ecology, monitoring, population and community genomics, and formalized collaborative efforts within and among diverse organizations will allow for a comprehensive understanding of genetic, ecological, and evolutionary relationships among the biological components of ecosystems.

Forest Service R&D will apply a holistic, multidisciplinary approach to studying invasive species at multiple ecological, spatial, and temporal scales that enhances our ability to provide synthesis for policymakers, communicate to diverse stakeholders, and use transparent processes to reduce conflicts.

Better understanding of interactions among multiple invasive species, native species, and disturbances will improve our ability to predict the establishment, spread, and effects of invasive species to more effectively mitigate negative socioeconomic and ecological outcomes or to capitalize on changed systems.

Predicting and Prioritizing Invasive Species

Forest Service R&D needs prioritization tools to focus research efforts on the most important invasive species. In addition, land owners and managers need decision support tools and prioritization models that help determine if, when, and where to apply specific measures for invasive species that pose the greatest threats to ecosystem functions, value, and services. Such decisions must take into account conflicting resource values, dynamic landscape patterns under varying management scenarios, and climate projections. Increased rates of globalization, the lack of strong networks across agencies and countries, and limited taxonomic expertise impede our ability to predict which species will be invasive and to detect, quickly identify, and respond to invasive species. Primary areas of research include the following:

- Developing science-based protocols to prioritize invasive species for research and management action.

- Developing a methodology for predicting which species are likely to become invasive.

Understanding the economic, ecological, and social effects of invasive species, which is incorporated in priority one, is also critical to prioritizing species and geographic areas for study and treatment.

Anticipated Products

Over the next 20 years, we anticipate concentrated efforts to support management goals that will result in the following:

- Models that provide a framework for ranking potential decisions across all taxa relative to their effect on resource values.

- Syntheses of existing science for species of current concern useful for identifying thresholds, dispersal patterns and pathways, and potential impacts and for monitoring protocols.

- Risk assessment models based on factors such as the probability of invasion and establishment and the effect on ecosystem goods, services, and values. Decision support tools for prioritizing management response and predicting outcome likelihoods.

- Tools and strategies to facilitate rapid communication of information on species occurrences and changing predicted invasiveness.

- Interactive Web interfaces to facilitate the delivery of models and tools.

- Models that identify new potential invaders based on knowledge of similar species and other variables such as climate and soils in the country of origin.

- Decision support tools for determining when and where to most effectively apply or stop treatment and control and for selecting the most appropriate measures in a particular circumstance.

Outcomes

Expected outcomes have been divided into a rough timeline with the following expectations:

Now

Enhanced international networks along with synthesis and rapid dissemination of information will improve risk assessment accuracy and enhance prevention and EDRR activities.

Recognizing the importance of structuring decision support tools to account for conflicting values will contribute to the goal of reducing future appeals and litigation.

Soon

We will predict likely effects of natural and anthropogenic disturbance on potential of key invasive species to establish, spread, and affect native species and ecosystems.

Multidisciplinary and multiagency teams will work together to develop integrated prioritization tools that optimize use of limited resources.

Decisionmakers will have a better understanding of the efficacy and costs of available treatment and control measures and have decision support tools to begin prioritizing when, where, and how they treat invasive species.

Virtual centers of excellence will share expertise across stations. SWAT teams—special weapons and tactics teams— that include Forest Service research will treat invasive species of the highest priority.

Future

Forest Service R&D will provide the science that enables multiagency collaborative efforts to address high-priority invasive species issues more quickly and vigorously. R&D has greater accountability for using science-based protocols for assessing high-priority issues and locations.

Tools that account for the role of disturbances, management actions, and climate scenarios will enhance the manager's ability to anticipate high-priority treatment areas of the future and, therefore, to make better use of limited resources to effectively manage and mitigate invasive species. A transparent and science-based process will enable managers to deal with conflicting values, reduce delay due to protracted disagreements, and enhance agency performance and accountability.

The Forest Service research results will improve credibility of management decisions because we will be better able to identify a number of species that are likely to be invasive, disseminate this information, and provide examples of how our assessment led to the curtailment of what could have been large invasive species outbreaks.

Identifying and Detecting Invasive Species

The capacity to detect, identify, and monitor species and variants of current and future invasive species from all taxa groups is critical. Prevention relies on rapid, reliable identification of potential or known invasive species entering North America or dispersing to new habitats within North America. Early detection of incipient invasions may make eradication economically feasible. Managers must be able to detect and delimit areas where invasive species are present to determine if, what, when, and where control measures should be applied. Primary areas of research include the following:

- Improving invasive species detection and diagnostics technology.
- Quantifying invasive patterns and processes across geographical and elevational gradients.

Taxonomy plays a major role in developing detection and diagnostic technologies. Understanding the phylogeographic relationships of current and potential invasive species can aid in identifying its country of origin, knowing what species to include in diagnostic keys and tests, and finding similar species that may help in developing management tools.

Anticipated Products

Over the next 20 years, we anticipate concentrated efforts to support management goals that will result in the following:

- Molecular markers developed to distinguish all stages of high-priority invasive pathogen and insect species or biotypes of concern and deployed in "high throughput"

detection tools (e.g., microarrays, real-time polymerase chain reactions, monoclonal antibodies).

- Models that identify new potential invaders based on knowledge of similar species and other variables such as climate and soils in the country of origin and United States, disturbance regimes, and other ecosystem characteristics.
- Synthesized scientific information for early detection protocols and survey methods.
- New and improved traps to detect presence and abundance of key species.
- Detection methods for organisms or their byproducts in wood or wood products.
- Remote sensing techniques to detect invasive damage and stress over large areas.
- Risk-based models to direct the search for invasive species to most likely habitats and locations.
- New and improved monitoring tools to detect and quantify new invaders and determine primary routes of spread.

Outcomes

Expected outcomes have been divided into a rough timeline with the following expectations:

Now

Through improved interagency and international cooperation, we identify and confirm pathways and targets and we develop integrated and innovative detection and diagnostic tools to prevent the introduction of new invasive species and reduce the spread of established ones into new areas.

Improved access to regional quarantine facilities owned or maintained in partnership by the Forest Service, other Federal agencies, universities, and international agencies enhances our ability to manage and mitigate invasive species.

We use Experimental Forests and Ranges, as well as Research Natural Areas, to understand invasiveness and the effects of climate change on invasive species.

Soon

Interagency communication identifies areas in which agencies can help each other cover taxonomy of all taxa. Improved taxonomic coverage will improve our ability to predict and prevent the introduction of many invasive species.

We will determine best methods for accumulating and disseminating a database of knowledge on key and high-priority invasive species.

We will deploy a mechanism to automatically trigger rapid threat assessment for new introductions so that we can minimize response, control, and restoration time.

Future

We will predict likely effects of climate change on invasive species.

Through increased data sharing and international collaboration, we will know nearly all species that are introduced and will be able to distinguish them from species already present in the habitat.

Managing Invasive Species and Altered Systems

In the broad sense, our desired resource outcome is that forest and range ecosystems are healthy and productive and that they provide a sustainable supply of services, products, and values that enhance the quality of life for present and future generations. Societal needs related to managing forests and rangelands under the influence of invasive species can be met only through quantifying and projecting system behavior and value under different scenarios. To formulate effective management policy at varying time and space scales, Forest Service R&D must provide probabilistic statements on the magnitude and direction of change; likely outcomes without intervention; options for rational action, including their costs; and systems and practices for accomplishing the rational actions. Research priorities in this area include the following:

- Developing more effective treatments, control measures, and management options for high-priority species and systems.

- Predicting interactions among multiple invasive species and multiple disturbances under varying climatic scenarios.

Anticipated Products

Over the next 20 years, we anticipate concentrated efforts to support management goals that will result in the following:

- Quantified efficacy and side effects of existing and new treatments and control measures.

- Expanded options for treatment and control that use biologically based technology for invasive species.

- Management strategies, methods, and practices for managing altered systems to continue to deliver needed goods, services, and values.

- Science-based decision support systems to help choose effective management actions at species and system levels.

- Decision support tools for determining when and where to most effectively apply or stop control efforts and for selecting the most appropriate measures in a particular circumstance.

- Management options to increase ecosystem resiliency and resistance to invasive species and to enhance natural recovery of affected ecosystems.

Outcomes

Expected outcomes have been divided into a rough timeline with the following expectations:

Now

Using existing and emerging information, we have developed and communicated options and costs for control and mitigation actions.

Forest Service R&D capitalizes on existing and newly established partnerships to synthesize existing knowledge regarding control measures for invasive species and to target research efforts at advancing control measures for the highest priority species.

Improved access to regional quarantine facilities owned or maintained in partnership by the Forest Service, other Federal agencies, universities, and international agencies enhances our ability to manage and mitigate invasive species.

Soon

Decision support tools will help managers identify thresholds that trigger specific management actions.

We will make science-based tools that provide tradeoff analyses of conflicting values and illustrate options and potential outcomes in probabilistic terms available to decisionmakers.

We will identify options for rehabilitating ecosystems or capitalizing on invaded systems.

Decision support tools and management options enable functional restoration of economically and ecologically critical systems.

Future

Through improved control measures and decision support models, land managers will safely and effectively manage and mitigate newly arrived and established invasive species and minimize their spread.

Science-based decision support systems will be used to help ensure continued provision of goods, services, and values from altered systems.

We will be able to identify when to rehabilitate an ecosystem or when to capitalize on the invaded system.

Skills Needed

A broad range of knowledge and skills is necessary to successfully accomplish these high-priority research objectives. The ability to work in and coordinate interdisciplinary teams within and across agencies is critical for many aspects of this research. Needed knowledge and skills include the following:

- Taxonomic expertise (insects, microorganisms, plants, fish, amphibians, reptiles)
- Systematics
- Phylogeography
- Genetics (classical, population, molecular)
- Ecology (chemical, population, invasion and disturbance, landscape)
- Conservation biology
- Basic biology (all taxonomic groups)
- Silviculture
- Forest operations
- Integrated pest management
- Statistics
- Toxicology and environmental fate
- Risk assessment and risk management

- Economics (ecosystem services, cost-benefit analysis, human values assessment)
- Sociology
- Geographic Information System mapping; modeling and analysis
- Computer database construction and mining
- Decisionmaking and conflict resolution
- Outreach; marketing and technology transfer
- Web and Internet programming

Process Check

Forest Service R&D periodically reviews and reevaluates its invasive species research to ensure responsive, timely, cost-effective delivery of science and products. This review occurs approximately every 5 years. Key questions to address during this review include the following:

- Are we working on the most important targets?
- Have we identified any new avenues to shore up taxonomic skills?
- Have we enhanced funding to address the four priority areas?
- Have we enhanced efforts to synthesize and disseminate information on species that have a higher potential of being invasive?
- Have we developed and tested a framework that enables us to assess across taxa what species and locations are priorities for research?
- Are we publishing research and synthesis papers in peer-reviewed outlets?
- Are we synthesizing available understanding in decision-support models?
- Are results being transferred to and applied by managers?
- Are we using our partnerships effectively?
- What new tools and technologies need to be incorporated into our approach?

Literature Cited

National Invasive Species Council. 2008. 2008-2012 national invasive species management plan. Washington, DC: Department of the Interior, Office of the Secretary, National Invasive Species Council. 35 p. http://www.invasivespeciesinfo. gov/council/mp2008.pdf (15 December 2009).

U.S. Department of Agriculture (USDA) Forest Service. 2003. Invasive species research and development logic model. Washington, DC: U.S. Department of Agriculture, Forest Service, Research and Development. 4 p.

Additional Reading

U.S. Department of Agriculture (USDA) Forest Service. 2001. 2000 RPA assessment of forest and range lands. FS-687. Washington, DC: U.S. Department of Agriculture, Forest Service. 78 p. http://www.fs.fed.us/research/rpa/ (15 December 2009).

USDA Forest Service. 2004a. Invasive species program and activity credibility through accountability business plan. Washington, DC: U.S. Department of Agriculture, Forest Service, Research and Development. 12 p.

USDA Forest Service. 2004b. National strategy and implementation plan for invasive species management. FS-805. Washington, DC: U.S. Department of Agriculture, Forest Service. http://www. fs.fed.us//invasivespecies/index.shtml (15 December 2009).

USDA Forest Service. 2005. Invasive species integrated goals strategy assessment. Washington, DC: U.S. Department of Agriculture, Forest Service, Research and Development. 1 p.

USDA Forest Service. 2006. Forest Service Research and Development Strategic Program Area review team handbook: a guide for review for team members and FS R&D staff. Washington, DC: U.S. Department of Agriculture, Forest Service, Research and Development. 32 p.

USDA Forest Service. 2007. USDA Forest Service strategic plan FY 2007-2012. FS-880. Washington, DC: Forest Service. 40 p. http://www.fs.fed.us/publications/strategic/fs-sp-fy07-12.pdf. (December 2009).

Forest Service R&D—Invasive Insects: Visions for the Future

Kier D. Klepzig[1], Therese M. Poland[2], Nancy E. Gillette[3], Robert A. Haack[2], Melody A. Keena[4], Daniel R. Miller[5], Michael E. Montgomery[4], Steven J. Seybold[6], and Patrick C. Tobin[7]

Abstract

The Forest Service has identified invasive species as one of four significant threats to our Nation's forest and rangeland ecosystems and likened the problem to a "catastrophic wildfire in slow motion." Forest Service Research and Development (R&D) has a crucial role in providing insight and options to protect trees, forests, and ecosystems from the threat of invasive insects. Currently, Forest Service R&D, in close coop-eration with Forest Health Protection, provides information crucial to the development of tools for studying, controlling, and mitigating several invasive insects in the United States, such as the Asian longhorned beetle, emerald ash borer, banded elm bark beetle, Mediterranean pine engraver, Sirex woodwasp, and hemlock woolly adelgid. Forest Service R&D also strives for a broad, principle-based framework applicable to current and future invasive insect problems. The historic breadth and depth of R&D charges this branch of the Forest Service with playing a leadership role in providing long-term, comprehensive, cross-cutting solutions. Clients and scientists have worked together to identify short- and long-term needs to enhance existing research. Examples of this vision are provided in this paper.

This visionary white paper outlines several specific needs derived from a review of future needs and the strengths and weaknesses in existing programs. These issues are discussed and prioritized under four headings: (1) Prevention and Prediction, (2) Detection and Eradication, (3) Management and Mitigation, and (4) Restoration and Rehabilitation. A special concern has been raised that the Forest Service is losing its capacity to provide biologically based technologies, such as pheromones, biological controls, microbial pesticides, and other environmentally sound mitigation options, unless existing and anticipated gaps in expertise are filled.

The Role of Forest Service Research and Development

Forest Service Research and Development (R&D) has the strategic capability to conduct research on the biology, ecology, and management of invasive insects on a national scale. The Forest Service National Strategy and Implementation Plan for Invasive Species Management (NSIPISM), developed by a multidisciplinary team of specialists, managers, and scientists, serves as one perspective on customer needs of any Forest Service invasive insects program. Specifically, this document states that "...the Forest Service is well positioned to be a leader nationwide and worldwide in the battle against invasive species. Our challenge is to learn to lead collaboratively."

Specifically, Forest Service R&D has the following capabilities:

- Broad existing authorities and responsibilities assigned by the Chief of the Forest Service.

- Expertise in research on land management, entomology, pathology, ecology, and several other specialties.

- National and international presence.

- Relationships with every State and territorial agency with responsibility for invasive species.

---

[1] Assistant Director, Forest Service, Southern Research Station, 200 W.T. Weaver Blvd., Asheville, NC 28804.

[2] Research Entomologist, Forest Service, Northern Research Station, Stephen S. Nisbet Bldg., 1407 S. Harrison Rd., Room 220, East Lansing, MI 48823.

[3] Research Entomologist, Forest Service, Pacific Southwest Research Station, 800 Buchanan St., West Annex Bldg., Albany, CA 94710–0011.

[4] Research Entomologist, Forest Service, Northern Research Station, 51 Mill Pond Rd., Hamden, CT 06514.

[5] Research Entomologist, Forest Service, Southern Research Station, 320 Green St., Athens, GA 30602–2044.

[6] Research Entomologist, Forest Service, Pacific Southwest Research Station, 1731 Research Park Dr., Davis, CA 95618.

[7] Research Entomologist, Forest Service, Northern Research Station, 180 Canfield St., Morgantown, WV 26505.

Strengths and Weaknesses of Forest Service R&D

Forest Service R&D is positioned to address the challenges posed by invasive species to U.S. forest resources. Applicable overall strengths of Forest Service R&D include the following:

- Breadth of skill sets among scientist ranks.

  - Molecular biology.

  - Chemical ecology.

  - Quantitative ecology.

  - Risk assessment.

  - Development of detection tools.

  - Development of management techniques.

- Access to a network of long-term sites.

- Ability to conduct longer term research.

- National presence and strategic plan.

- Ability of geographically dispersed units to address regional and national problems and coordinate across regions.

- Extensive infrastructure (labs, quarantine facilities, experimental forests).

- Ability to determine consequences and devise mitigation strategies for destructive invasive forest insects in advance of their establishment.

- Ability to discover and rear natural enemies, determine their nontarget effects, release them, get them established, and monitor their effect on the target pest and nontarget species.

- Unique partnerships with the National Forest System (NFS) and State and Private Forestry (S&PF), providing access to sites, technology transfer, and management expertise.

- Focus on forest pests.

- Focus on research rather than regulatory aspects.

To adequately deal with future needs around invasive species research, Forest Service R&D must also attempt to address weaknesses, including the following:

- A need for research-quality taxonomic expertise, including morphology, chemotaxonomy, and molecular taxonomy. The cooperation between Forest Service R&D and the Forest Health Protection (FHP) program in training taxonomists for use in the Early Detection and Rapid Response Program is an example of preliminary efforts to address this need.

- The need to strenghten in-house capabilities in molecular genetics and epigenetics.

- The need to integrate skill sets within Forest Service R&D to better address complex, multidisciplinary problems.

The Identity and Needs of Our Customers

In invasive species research, existing (and potential) customers are diverse, including the USDA Animal and Plant Health Inspection Service (APHIS), NFS, FHP, State agencies, and all entities required to have published sources to support their decisions. For these customers, as well as the academic community, publications in peer-reviewed scientific journals remain the standard products. Peer-reviewed competitive requests for proposals have been successful in drawing these customers together into cooperative research on a targeted topic. Careful consideration (with comprehensive user group input) of critical Forest Service R&D needs, however, is essential to credibility, relevance, and buy-in from users and supporters of such a program.

In addition, Forest Service R&D must respond to the technology transfer needs of forest land owners and managers, urban foresters, arborists, nursery and timber industry representatives, nongovernmental organizations (NGOs), Timberland Investment Management Organizations, and Real Estate Investment Trusts. Traditional products (refereed publications, scientific meetings) may not be sufficient to meet their needs. Partnerships with the FHP program, universities, cooperative extension, and NGOs may be especially valuable in this outreach. Partnerships through Forest Service International Programs, as well as international grants, may be useful in cooperating with international collaborators dealing with similar pests.

The development of true research partnerships (including joint study-plans, personnel sharing, sharing of administrative burdens, joint publication through Forest Service and external outlets, and joint presentations) is essential to success in this arena.

Responding to the Invasive Species Strategic Program Area Peer Review Panel

The recent peer review of the Invasive Species Strategic Program Area (SPA) provides some insight for the future of invasive species research within the Forest Service. The specific suggestions for improvement, development, and growth are listed in the following text.

Enhance External Partnerships

The need to focus more on interdisciplinary, cooperative research was clearly identified, although an overemphasis on outside cooperators could lead to Forest Service R&D becoming a funding organization with no ability to conduct research. Bilateral reciprocity agreements are one possible vehicle for addressing this need. The biggest problem in studying nonindigenous insects is the paucity of populations at the beginning of an invasion. One solution would be for scientists in the United States to work on species that are of concern to other countries while their scientists work on species that are of concern to us. Scientists in each country would have a high likelihood of enhancing our knowledge and developing effective solutions. Restricting the number of participating countries (e.g., China, Korea, Sweden, Canada) would ensure balanced reciprocity. Scientists could then justify their work-in-country under the agreement. We also enjoy and employ a broad array of international contact networks (e.g., Chinese Academies of Science and Forestry, International Union of Forest Research Organizations, members of many international pest quarantine research groups). This network provides resources, contacts, and much-needed background information on new species.

Improve Integration Within Federal Agencies

As a Federal Government agency within USDA Forest Service R&D has close administrative and cultural ties with APHIS, the USDA Agricultural Research Service (ARS), the U.S. Department of Homeland Security (within Customs and Border Protection), and other Government agencies. These ties facilitate the sharing of resources and research sites (e.g., ports of entry). Nevertheless, interagency integration could be improved to include better communication, coordination, and definition of roles. In some cases, APHIS is the lead agency in regulating and managing invasive species, and Forest Service FHP has the lead in others; however, neither is a research organization. Forest Service R&D and ARS are research entities engaged to find scientific solutions to invasive species problems. Clearly defined roles and improved communication can lead to better coordination and less competition among Federal agencies.

Improve Communication of Research Results

Communication of research results could be improved by strengthening internal networks (with FHP, S&PF, and NFS), developing a team of communication experts to identify key users and evaluate the most effective methods to reach them, increasing effective use of university extension services, partnering with end users in Forest Service R&D projects, and increasing participation in technical programs. An especially important step will be to engage end users, as much as possible, in research programs from the start in order to facilitate technology transfer and ensure operational compatibility. This approach could include something as technical as systems engineering or even a formal process such as that used in ARS.

Maintain or Enhance Infrastructure

One strength of Forest Service R&D is our extensive infrastructure that provides the facilities and capability to lead a national invasive species research program. To deliver successful programs, it is imperative that this infrastructure—particularly experimental forests, quarantine and rearing facilities, and collections and databases for long-term studies—be maintained and enhanced. Taxonomic expertise is in short supply in many disciplines; this is especially critical for invasive species research. The need to strengthen the interface between molecular technology and invasive insect science can be enhanced inhouse and through partnerships between Forest Service scientists and molecular biologists at universities. Past successes in documenting the source of invasions and connecting genetics to behavior have arisen from this model (e.g., Asian gypsy moth, hemlock woolly adelgid).

Short- and Long-Term Goals

In addressing the recommendations of the SPA review, it is useful to identify achievable, quantifiable goals. By necessity, some of these goals will be realistically set for the short term. Larger strategic goals will capitalize on the ability of the Forest Service to conduct long-term research.

Prevention and Prediction

Prevention and Prediction—Short Term

This goal ties in with broader priorities (identified in the Invasive Species SPA paper, "Overarching Priorities"): Predicting and Prioritizing, and Managing Invasive Species and Altered Systems. The NSIPISM also identifies the following short-term priority issues within the area of prevention:

- Work with APHIS and other partners to conduct analyses on invasion pathways and species risk assessments so that we can identify priorities and develop regulatory response plans.

- Build awareness of invasive species and their threat at all levels and jurisdictions.
- Complete a national research risk assessment to identify high-priority invasive species and continue working in their countries of origin to develop techniques to deal with these high-priority pests before they are introduced into the United States.

Other priority issues include the following:

- Need for more emphasis on developing background information on potential invasive species (e.g., forest insects of China) in English.
- Improved measures to reduce the risks of arrival through pathways such as solid-wood packing material, nursery stock, and other components of international trade and travel.
- Significant advances in the efficacy of fumigation and heat treatments for control of invasive species in solid-wood packing materials that arrive in the United States by sea and air.
- Development of new treatments in support of implementing higher standards for the bilateral shipment of these materials between North America and trading partners in Eurasia.

We also need to develop database systems (both nationally and internationally) to facilitate the transfer of information for intercepted species and prospective invaders. These systems should provide easy online keys and photos of all the stages. We need clear guidelines on acceptable risks and interception thresholds. We need to gather information on known invasive species affecting forests in other world areas and develop quick and cost-effective tools to identify intercepted species, especially because systematic resources are limited. This approach could minimize costs associated with eradication and management and could mitigate impacts to our Nation's forest and rangeland ecosystems.

To be able to create and manage resilient ecosystems, we must increase our understanding of the invasion potential of nonindigenous insects (i.e., species invasiveness) and the habitat characteristics that increase or decrease the ability for a new invader to establish (i.e., habitat invasibility). Understanding invasiveness is critical for improving our ability to predict the threat of nonindigenous species rather than relying on pest status in their native habitat.

Prevention and Prediction—Long Term

Any attempt to understand longer term scenarios involving prevention (and for that matter all other aspects of long-term invasive species research and development) will require, of necessity, predictive models. Models simulating long-term dynamics of invasive species will likewise need to involve consideration of changing forest ecosystems. Climate change has the potential to exacerbate insect outbreak intensity, increase voltinism, and permit the colonization of new habitats previously not susceptible to invasion. Fragmentation and other anthropogenic forces have the potential to drastically alter the movement of distribution of invasive species. The need for these long-term models will likely highlight the need for basic biological and ecological information that will inform their accuracy and utility.

Detection and Eradication

Detection and Eradication—Short Term

This goal ties in with Overarching Priorities 3 (Identifying and Detecting) and 4 (Managing Invasive Species and Altered Systems). NSIPISM also identifies the following short-term priorities for early detection:

- Establish a Forest Service-wide early detection and rapid response emergency fund. Develop guidelines to ensure that funds are immediately available to respond to newly discovered alien species.
- In partnership with the Forest Health Technology Enterprise Team, develop maps of priority ecosystems and habitats placed at risk from invasive species.
- Working with partners, develop rapid response incident teams that cross jurisdictional lines and respond quickly to newly established populations of invasive species.
- In partnership with the U.S. Departments of Homeland Security, the Interior, and Agriculture; State agencies; and others, develop high-speed, reliable, and robust technologies to detect and respond to introduced invasive species.

Other identified Detection and Eradication needs:

- Cooperative efforts with China and the European Union to develop a series of "sentinel forests" surrounding ports of entry to facilitate early detection.
- Development of improved cost estimates of impacts to inform decisions and regulations about international trade, eradication, and suppression efforts.

- Enlistment of public support from birdwatchers, master gardeners, arborists, tree-care professionals, warehouse workers, and others to look for unusual evidence of insect activity.

- Improved understanding of the relative efficacy of eradication tools for different types of invasive insects (e.g., Bt sprays for Lepidoptera [moths], tree removal for woodborers).

- Decision models to provide guidance in implementing eradication measures.

- Development of survey and trapping programs for broad categories of insects.

- Increased research and development involvement in the urban forests of the country.

- Seeking of opportunities to communicate with arborists, golf course managers, urban parks personnel, and city and county land managers about the invasive species problem.

- Elevation of priority of urban areas for future survey and detection efforts.

Detection and Eradication—Long Term

We need to develop new detection technologies, such as olfactory, acoustic, x-ray, ultrasound, thermal, and infrared. A critical component of detection programs is the development of cost-effective delimitation strategies employing lower cost detection tools that can be deployed at the landscape level. We need to develop new eradication techniques, such as the effective use of pheromones and other semiochemicals, to direct or target our eradication efforts in the future. We need to strengthen our collaborations with State agriculture officers and APHIS to facilitate eradication efforts.

Management and Mitigation

Management and Mitigation—Short Term

This priority ties in with Overarching Priority 4 (Managing Invasive Species and Altered Systems). Short-term needs are also identified in the NSIPISM as "Control and Management":

- Complete the comprehensive (all invasive species) inventory and mapping for all national forest land and water, including neighboring land, where appropriate.

- Conduct a comprehensive (all invasive species) risk assessment based on existing information for the specific purpose of identifying priority species and areas for program focus.

- Through research and other means, develop additional tools, such as biological, cultural, chemical, and physical controls, for priority species. Identify mechanisms involved in the arrival of these priority species and their successful establishment.

- Monitor long-term invasive species population trends and measures of treatment efficacy. Make this information readily available to all stakeholders, public and private.

Conducting research into new methods of controlling the spread of, or even the eradication of, isolated populations could yield significant advances. Economic analyses of impacts may aid in prioritizing investment in invasive species research, development, and application.

We need improved cost estimates on the type and extent of treatments that are needed in eradication and management. Chemical controls are often quick and inexpensive in the short run but may have undesirable and unintended nontarget effects. In the case of biological control agents, we must adequately evaluate their ability to establish and suppress populations of the targeted invasive insect.

Management and Mitigation—Long Term

Expanding classical biological control research would require developing primary quarantine and rearing facilities and delivering necessary resources to the units that maintain them.

Restoration and Rehabilitation

Restoration and Rehabilitation—Short Term

This goal ties in with Overarching Priority 4 (Managing Invasive Species and Altered Systems). NSIPISM also identifies several short-term priorities:

- Prioritize and develop native plant stock that is resistant to invasive insects and pathogens. Although this goal is identified in the NSIPISM as short term, it is normally a long-term effort.

- Work internally and externally to identify budget and capacity for implementing the national strategy.

- Establish multidisciplinary invasive species management teams in each region/station to implement the national strategy and implementation plan.

- Update and enhance the Forest Service's invasive species Web site to serve as a comprehensive communication tool.

- Work with partners to (1) develop a targeted marketing strategy to achieve public awareness of invasive species and an understanding of the role citizens can play; (2) complete the invasive species best management practices video series and handbook; (3) expand quarantine facilities for plant, insect, and pathogen control research; and (4) increase the availability of taxonomists to identify new invasive species.
- Work with other agencies, such as the USDA Economic Research Service, to expand economic impact assessments for priority invasive species.

We need to develop prescriptions for habitats and landscapes (urban and natural) that are resistant or resilient to invasion.

Restoration and Rehabilitation—Long Term

We must develop an understanding of mechanisms of tree resistance to invasive insects that are responsible for huge losses of specific tree species. In general, this research is very long term, requiring collaboration among entomologists, molecular geneticists, and tree physiologists. The development of transgenic trees resistant to invasive insects is also a long-term possibility.

Opportunities and Lessons Learned

Opportunities for Increased Efficiency and Cooperation

Because limited resources currently are invested in the area of invasive species research, very little duplication exists among stations and especially within invasive insect research. Nevertheless, some duplication (with communication to ensure complementary efforts) is desirable as individual labs and scientists working on similar problems often use different techniques and approaches. In addition, because the impacts of many invasive insects are national in scope, the scientists working on them frequently conduct research at a broad geographic level. Connections among scientists working on similar problems already exist, but they could be strengthened.

The core areas to maintain and enhance are basic biology, ecology (i.e., landscape, population, chemical), and biocontrol. We must consider Forest Service R&D's current and projected capacity for work on invasive insects, especially our dimin-

ished capacity for taxonomy among the research science programs nationally. It would also be helpful to have a database or library of detection baits for invasive species, particularly bark beetles for which we have some chemical ecology background. It is especially important to increase our taxonomic capacity, which would provide long-term institutional memory and structure for detecting nonindigenous species and for understanding and distinguishing them from native ones.

The need for entomologists trained in developing biologically based technologies for use in control programs is acute. Currently entomologists who have little direct training or are near retirement are conducting most of this work. The expertise concentrated in the Northern and Southern research stations should help leverage dollars and maintain critical mass in this area of inquiry, but it is crucial to develop a more geographically effective distribution of resources to face this issue.

Forest Service R&D must think broadly and be flexible. The appropriate mix of basic and applied research will be especially crucial. Some highly specialized basic research is required to develop critically needed biological and ecological knowledge bases; however, a critical need also exists for research with broad applications and the ability to shift among taxa and ecosystems. In particular, mechanistic-based research may tend to have especially broad implications and applications. In most cases, by the time we detect new invasive species, they are already established. Because many invasive species in the United States are not economically important in their native habitat, very little previous research is available; hence, we often are faced with the need to answer even the most fundamental research questions. To be truly effective, Forest Service R&D must partner with FHP, NFS, USDA APHIS, State agencies, and universities. We also need a certain degree of hyperactivity in outreach, such as pest alerts and forest insect and disease leaflets, university cooperative extension publications, presentations at work conferences, and arborists meetings.

Emerging Opportunities in the Arena of Invasive Insects

Invasion pathways and risk assessments are emerging as crucial fields in the study and management of invasive species. Understanding pathways (wood material, plants for planting, other commodities) is being emphasized as a promising approach in preventing the establishment of new invasive species. For example, recent studies have highlighted the applicability of gravity models to facilitate our understanding of the human-mediated interactions between ports of egress and ports of entry that are critical in quantitatively assessing invasion pathways. Another emerging and increasingly rigorous field is that of risk assessment, which considers the risk of a specific invasive species, the magnitude of the potential loss from the risk, and the probability that such a risk will occur. Both of these areas are being emphasized in the creation and support of threat assessment centers.

A particularly worrisome—and probably under addressed—emerging problem is that of the acquisition of exotic fungi by native insects and the acquisition of native fungi by exotic insects and of the risks that these new relationships pose to native forests. The consequences are unforeseeable, but beetles are great disseminators of fungi. The need for DNA databases for fungal associates of beetles worldwide is thus paramount and would aid in our understanding of, for example, the impacts of ambrosia beetles and the mortality they may cause. Forest Service R&D should assume a leading role in ongoing international efforts to establish and coordinate global databases of molecular tools for species identification (e.g., Consortium for the Barcode of Life [CBOL] and GenBank®). Forest Service R&D should lead efforts to incorporate fungal associates of beetles and forest pests into the All-Fungi Initiative of CBOL. It is imperative that all new molecular sequences are linked with voucher specimens to ensure validity of the species name associated with the sequence. This effort requires the collaboration of taxonomists and molecular biologists. Forest Service R&D should lead the required phylogenetic work to support molecular databases. It is critical to develop molecular markers for species identification, especially for immature stages of native and commonly intercepted exotics.

A real opportunity exists to research ways to manipulate the Allee effect (wherein, in smaller populations, the reproduction and survival of individuals decrease, with the effect usually disappearing as populations grow larger) to facilitate management. Certain processes may lead to a decline in the population of an invasive species with a decline in its density; for example, the use of tactics designed to disrupt mating (mass trapping, release of sterile insects). Research that focuses on the manipulation of Allee dynamics may be especially useful and needed.

At present, we have numerous species of nonindigenous insects in the United States, and many more likely are not yet known to us. Even for those insects that we have detected, however, we lack an understanding of their potential impacts in our forest ecosystems. We especially need to initiate long-term research projects on the impacts of exotic insects on our native ecosystems. Such work will give us a much better idea of the magnitude of their costs (and maybe benefits), thereby providing a comprehensive base to support future research efforts. Long-term studies that record the presence of invasive species across robust spatial and temporal scales can also facilitate the development of models of spread and of the factors and processes that limit or enhance spread rates. This research would provide insight into climatic or biological factors affecting the spread of invasive species and would provide much needed data for risk analysis and policy decisions. Species-specific models could in turn be used to develop broader paradigms of spread applicable to other systems and invaders.

Other opportunities/visions provided by cooperators and customers include the following:

- Establishing a Forest Service-wide emergency fund and technical advisory teams to provide rapid response to new threats of invasive species.

- Increasing the efficacy of research partnerships and collaboration through more stable funding and through the monitoring and measurement of results.

- Establishing Forest Service-wide service centers funded by and overseen by two or more stations to provide mapping, quarantine facilities, molecular technology, etc.

- Increasing research efforts both internationally and nationally to identify ecosystem processes that provide resiliency to invasive species.

- Developing scientifically sound treatments for solid-wood packing material.

- Increasing our ability to detect and predict ecosystem change by better integration of research plots and information with Forest Inventory and Analysis databases.

- Developing improved estimates of the socioeconomic impacts of invasive insects, incorporating improved estimates of the costs and benefits associated with invasive species prevention and control, and collaborating with other agencies such as the USDA Economic Research Service.

Top Priorities

The direction from the U.S. Government Accountability Office and the Office of Management and Budget is that research has to be accountable and productive. With limited funds and people, the Forest Service R&D should maintain a balanced perspective in research endeavors. The effort should be high in short-term efforts that have a high likelihood of producing results. Therefore, efforts on any study should be based on the likelihood that such work will result in creating the science (adding to our knowledge base and developing new approaches), using the science (developing new/more effective tools), and furthering the science (increasing our understanding of risks and impacts).

One possible approach may be provided by considering the different types of pestiferous invasive insects. In this case, it is hypothesized that the likelihood of research creating, developing, and furthering science varies with insect type and with the different strategies of the invasive species strategic plan (table 1). In detecting new invasive species, fast and reliable detection methods are needed. The sooner a new pest is detected, the sooner an eradication program can be initiated, in general, over a smaller area. The feasibility and costs of eradication are directly related to the degree and extent of establishment. For example, lepidopteran defoliator pheromone lures that are effective even at low population densities have been highly effective in detecting newly established populations of gypsy moth that can be cost-effectively eradicated. When introduced in the past, the Asian gypsy moth has been successfully eradicated through the use of aerial sprays, ground sprays, and egg mass surveys because of our ability to detect low-density populations. Therefore, a high priority should be research aimed at developing extremely sensitive pheromone-baited traps for other invasive species. Forest Service R&D scientists have the expertise to tackle this objective.

In contrast with the understanding of pheromones for moths (*Lepidoptera*) and beetles (*Coleoptera*), pheromones are poorly understood for woodwasps (*Hymenoptera*) and for aphids and scales and their allies (*Hemiptera* and *Homoptera*). An early detection system for these invasive insect groups will likely involve either visual inspections or some new technology yet to be developed as opposed to pheromone-baited traps. It might likely be some time before publishable results would arise from such efforts. Work on detection for aphids and scales should be "low." Maintaining such work at a low level over a long period has a high likelihood of accomplishing some significant, but unpredictable, advances in the 25-to-30-year future, basically a "speculation" component to the investment strategy. At present, the best investments in dealing with introduced exotic aphids and scales are likely mitigation at a "medium" level with introductions of biological controls and restoration at a "high" level with a resistance breeding program. Another important component that is higher with aphids and scales would be prevention. Given the lack of ability to produce effective detection and eradication tools for these pests (in contrast with moth defoliators and bark beetles), it is better to find ways to minimize movement of such organisms from their country of origin. Certification programs at processing areas that ensure "Free of aphid or scale" would reduce the need to even deal with them. Similar approaches may be effective in prioritizing research and development regarding other invasive species.

Summary and Skills Needed

As stated in the Invasive Species SPA paper, "Overarching Priorities," "a holistic national strategy will improve sharing of expertise across research Stations, and encourage actions that prevent regional threats from expanding into national ones." We see the need for skills and skilled personnel to make these

Table 1.—*Research investment matrix: relative profitability of research for different groups of invasive insects and strategic goals.*

Strategic goal	Insect group			
	Defoliators	Bark beetles	Wood borers	Aphids/scales
Prevention	Low	Medium	Medium	High
Prediction	Medium	High	High	Medium
Detection	Very high	High	Medium	Low
Eradication	Very high	High	Medium	Low
Mitigation	Medium	High	High	Medium
Restoration	Low	Medium	Medium	High
Education	Medium	Low	High	High

visions a reality. Among the skill sets identified in the "Over-arching Priorities" paper, we see the following skills as being especially valuable in addressing Forest Service R&D needs in the area of invasive insects:

- Taxonomic expertise (especially insects and associated microorganisms).

- Systematics (morphological and molecular, insects and associated microorganisms).

- Genetics (classical, population, and molecular).

- Ecology (chemical, population, invasion/disturbance, and landscape).

- Basic biology (especially insects and associated microorganisms).

- Integrated pest management.

- Monitoring design.

- Toxicology and environmental fate (of pesticides).

- Risk assessment.

- Economics (impacts of outbreaks).

- Technology transfer (pest alerts, monitoring, and quarantines).

Invasive Forest Pathogens: Summary of Issues, Critical Needs, and Future Goals for Forest Service Research and Development

Ned B. Klopfenstein[1], Jennifer Juzwik[2], Michael E. Ostry[2], Mee-Sook Kim[3], Paul J. Zambino[4], Robert C. Venette[2], Bryce A. Richardson[1], John E. Lundquist[5], D. Jean Lodge[6], Jessie A. Glaeser[7], Susan J. Frankel[8], William J. Otrosina[9], Pauline Spaine[9], and Brian W. Geils[10]

Abstract

Invasive pathogens have caused immeasurable ecological and economic damage to forest ecosystems. Damage will undoubtedly increase over time due to increased introductions and evolution of invasive pathogens in concert with complex environmental disturbances, such as climate change. Forest Service Research and Development must fulfill critical roles and responsibilities to address issues related to invasive forest pathogens. This paper identifies critical, long-term research needs in four key areas: (1) prediction and prevention, (2) early detection and rapid response, (3) management and mitigation, and (4) restoration and rehabilitation. The paper also addresses issues related to national and international collaboration, scientific applications, and communication.

General Overview

In early 2007, diverse forestry professionals from the Forest Service and other institutions were surveyed to help determine the critical issues, needs, and top priorities of invasive pathogen research for Forest Service Research and Development (R&D). The consensus of the feedback received is reported in this section. Synthesized information is reported in more detail in the paper's subsequent sections.

The Issues

Ecosystem damage caused by invasive forest pathogens is often severe, long term, widespread, and difficult to mitigate. These pathogens affect ecosystems across forest landscapes and ownerships by reducing the viability of plant and animal species, decreasing forest productivity, and impairing carbon capture. Global climate change and increasing global trade and travel further escalate the threats from invasive forest pathogens. Continuing commitment and cooperation are critical to develop effective approaches for mitigating ecological, economic, and sociological effects and for managing healthy ecosystems that support human needs. Research to develop these approaches requires the integration of plant pathology with other multidisciplinary expertise, and frequently relies on long-term and/or multiscale studies. Success depends on maintaining and building public support and collaboration among State, Federal, and international forest health specialists and natural resource

[1] Research Plant Pathologist and Research Geneticist, respectively, Forest Service, Rocky Mountain Research Station, 1221 S. Main St., Moscow, ID 83843.

[2] Research Plant Pathologists, Forest Service, Northern Research Station, 1561 Lindig Ave., St. Paul, MN 55108.

[3] Assistant Professor, Kookmin University, Department of Forest Resources, 861-1 Jeongneung-Dong Seongbuk-Gu, Seoul, Korea 136–702.

[4] Plant Pathologist, Forest Service, State and Private Forestry, Forest Health Protection program, Region 5, Southern California Shared Service Area, San Bernardino National Forest—SO, 602 S. Tippecanoe Ave., San Bernardino, CA 92408.

[5] Supervisory Forest Entomologist, Forest Service, Pacific Northwest Research Station, Forest Health Protection program, 3301 C St., Suite 202, Anchorage, AK 99503.

[6] Botanist, Forest Service, Northern Research Station, P.O. Box 1377, Luquillo, PR 00773.

[7] Research Plant Pathologist, Forest Service, Northern Research Station, One Gifford Pinchot Dr., Madison, WI 53726.

[8] Sudden Oak Death Research Program Manager, Forest Service, Pacific Southwest Research Station, 800 Buchanon St., Albany, CA 94710.

[9] Research Plant Pathologists, Forest Service, Southern Research Station, Diseases and Invasive Plants Team, 320 Green St., Athens, GA 30602.

[10] Research Plant Pathologist, Forest Service, Rocky Mountain Research Station, 2500 S. Pine Knoll Dr., Flagstaff, AZ 86001.

managers. The role of plant pathology research is to acquire new knowledge and to develop and test techniques useful to public agencies, especially the U.S. Department of Agriculture (USDA) Forest Service Forest Health Protection (FHP) program and Forest Service National Forest System (NFS); USDA Animal and Plant Health Inspection Service (APHIS); and other State, national, and international management or advocacy groups. In response to the threats from invasive forest pathogens, a number of critical research needs are identified and several priorities are selected that describe gaps in information, analysis, synthesis, research scope, collaboration, and capacity.

Critical Needs

- The capacity to recognize and identify species and variants of current and future invasive pathogens, hosts, microorganisms, and vectors through the use of classical and molecular taxonomy, fungal and plant systematics, and molecular diagnostic tools.

- Multidisciplinary studies that address and provide attainable solutions to complex forest health issues (e.g., interactions between climate change, atmospheric chemistry, forest fragmentation, land-use change, silviculture, and wildfire and/or insect disturbance on invasive pathogen behavior in the Nation's rural and urban forests).

- Knowledge of ecological and genetic responses to determine the following:

  - Biophysical factors involved in pathogen spread, potential transport pathways, and changes after establishment.

  - Mechanisms of defensive host response to pathogen challenge that are heritable, preformed, or environmentally inducible.

  - Host populations that respond similarly or are ecologically valuable for genetic conservation efforts.

  - Relationships and coevolution among organisms in their historical and geographical context using genomic and phenomic approaches.

  - Potential development of novel species, hybrids, and subspecies with new ecological behavior.

- New tools and methods to predict, detect, and monitor potentially invasive forest pathogens, either prior to introduction or soon after establishment, through integrated collaboration with Forest Service FHP, Forest Service NFS, USDA APHIS, and national and international collaborators.

- Understanding the interactions of current or altered biotic and abiotic factors on the establishment, spread, and effects of invasive pathogens over time and space.

- Assessment and actionable knowledge of invasive pathogen costs to society through integrated economic and social science research with universities and other partners; such information is necessary to prioritize intervention responses (threat analysis and mitigation).

- Biological control and alternative-management techniques.

- Techniques to manage and rehabilitate forest ecosystems affected by pathogen invasions, and management practices that foster resilience in forest ecosystems before they are affected by invasive pathogens.

- Precise baseline information on worldwide distributions of forest pathogens (using molecular diagnostics), distribution of host populations, climate and niche data, and other environmental data through integrated collaboration with the Forest Service Forest Inventory and Analysis (FIA) program, Forest Health Monitoring (FHM) program, FHP, and international collaborators.

- Databases, Web sites, and better coordinated information sharing among forest health professionals as well as effective information exchange and educational outreach to other stakeholders.

Top Five Priorities

1. **Obtain baseline information on worldwide geographic distributions and environmental responses of forest pathogens, strains, hosts, vectors, and associated microbes.** This baseline information is critical to all invasive species program areas worldwide. Systematics expertise, supporting biological collections, and diagnostic tools developed by Forest Service R&D and university partners should be maintained and increased to support survey work (e.g., by FHM, USDA APHIS, and international collaborators). International collaborations are essential to predict, prevent, and detect invasive species introductions to the United States and other countries worldwide. Integrating Geographic Information System-based ground surveys, spatial modeling, and remote sensing can assess and predict organism distribution and environmental influences. Databases of collected information and technologies must be widely accessible for multiple purposes.

2. **Conduct climate modeling, risk assessments, and pathway analyses for priority hosts and pathogens.** Spatial biophysical data must be integrated at the appropriate scales for risk assessment of potential pathogen invasions or evolution of novel invasive pathogens. When combined with evaluations of host susceptibility to invasive pathogens, these assessments are fundamental for predicting and preventing future invasive pathogens, prioritizing the management of established invasive forest pathogens, rehabilitating affected ecosystems, and enhancing ecological resilience.

3. **Determine interactions of invasive pathogens with other disturbances.** Climate change, altered fire regimes, increases in insect damage, forest fragmentation, urbanization, floods, wind, and other human-natural ecosystem interactions will likely exacerbate effects of forest pathogens. Many forest diseases will become more widespread and damaging in changing environments.

4. **Implement long-term, ecosystem-based research to determine appropriate, adaptive management to restore the function and structure of affected ecosystems, enhance endemic biological control agents, identify genetic-based disease resistance for breeding and regeneration programs, foster natural recovery, and improve the resiliency of ecosystems to unexpected future threats.** Short-term research can provide much-needed progress; however, invasive forest pathogens also present many long-term issues that can only be addressed through long-term research.

5. **Build capacity for plant pathology research and associated collaborative efforts for responding to increasing effects of invasive forest pathogens.** The scope and complexity of the identified priority research present a serious challenge to Forest Service R&D within its present internal capacity and ability to leverage collaboration. Although such research requires collaborative, multidisciplinary effort, plant pathologists provide subject expertise and methods of study that are a core need for investigations of invasive forest pathogens.

Introduction

Invasive fungal pathogens have caused large-scale ecological and economic damage to forests in the United States (Lovett et al. 2006). Damage caused by these pathogens has been more severe, long term, widespread, and difficult to mitigate than that caused by any other biological disturbance agent. In the last century, pathogens introduced into our native forests have threatened the extinction of native tree species and critically degraded many diverse ecosystems across North America. Prominent forest diseases caused by invasive pathogens include chestnut blight (Anagnostakis 1987), Dutch elm disease (Brasier and Buck 2001), white pine blister rust (Geils et al. 2010), butternut canker (Furnier et al. 1999), dogwood anthracnose (Daughtrey et al. 1996), laurel wilt (Fraedrich et al. 2008), sudden oak death (Rizzo et al. 2005), pitch canker (Gordon et al. 2001), and Scleroderris canker (Hamelin et al. 1998). Ecological effects of invasive pathogens extend well beyond the affected trees (e.g., white pines, chestnut, elms, dogwood, butternut, oaks, and pines) to include other ecosystem components, such as dependent forest flora and fauna. Forest trees include keystone species providing food and structure for animal species (e.g., grizzly bears, small mammals, birds, etc.) and other plants. Invasive pathogens have severely diminished the productivity, sustainability, and ecosystem service functions of many forests. Damage from pathogens can also alter forest succession and forest species composition.

Invasive forest pathogens can originate from intercontinental or intracontinental movement, range expansion due to climate change, the creation of novel pathogens via evolution or hybridization, or other means (Palm 2001). The likelihood for invasive pathogen introduction will increase, however, as international trade and travel increase. The potential for invasive pathogen establishment and spread will also increase as newly introduced pathogens interact with changing climate, native insect pests and pathogens, wildfire, forest fragmentation, and other natural or human-induced disturbances. Equally challenging are threats from native invasive pathogens (Otrosina 2005). Changing climate, forest management, disturbance regimes, and increased potential for genetic hybridization among existing (and introduced) pathogen populations can initiate or exacerbate disease problems.

Forest Service Research Roles and Responses to Invasive Forest Pathogens

Throughout its history, the Forest Service has played a prominent role in responding to invasive forest pathogens. The following information briefly reviews the roles and responsibilities of the Forest Service R&D deputy area, as determined by congressional mandate, Executive order, institutional directives, and traditional practices.

Background Information

The R&D deputy area of the Forest Service has been mandated by Congress and by Executive order to conduct and facilitate research on invasive forest pathogens, to translate research results into actionable knowledge, and to communicate research results for application in management. In the United States, the Forest Service has a principal responsibility and capability to develop the necessary knowledge and tools for addressing invasive forest pathogen issues across State and regional boundaries. Furthermore, it builds and facilitates multidisciplinary, collaborative efforts among diverse research institutions and stakeholders to address national and international issues of invasive pathogens. It uses its unique and important leadership role to acquire information for the public interest and coordinate research programs to protect or enhance public and private natural resources in rural and urban settings. Because invasive pathogens invariably represent long-term issues, the Forest Service R&D has maintained primary responsibility of long-term research studies (e.g., permanent plots, experimental forests, biological collections, host genetics materials, and data archives) to address long-term issues associated with invasive pathogens. In addition, Forest Service R&D implements its mandated charter to conduct research and study national and international issues of invasive forest pathogens as they relate to the United States, to provide knowledge needed for protecting forests and other terrestrial ecosystems of the Nation.

Key Research Needs for Addressing Invasive Forest Pathogens

Several research needs must be addressed to maintain and enhance the Forest Service's capacity for effective response to invasive forest pathogens.

A strong systematics capability is essential to identify biotic components of disease caused by invasive pathogens. This capability is essential to obtain baseline data on the distribution of invasive pathogens. Specific examples of the needs for systematics capabilities include the following:

- Classical and molecular taxonomic expertise for the identification of invasive pathogens, hosts, associated microorganisms, and vectors.
- Phylogenetics expertise for elucidating evolutionary relationships within hosts, pathogens, and associated microbes (e.g., potential biocontrol agents).
- Advanced molecular characterization and molecular diagnostic tools for diagnostics and monitoring activities.
- Maintenance and expansion of fundamental systematics resources, such as herbaria, survey and collection records, culture collections, and DNA sequence databases.

Knowledge of ecological behavior and impacts of invasive pathogens worldwide is essential for assessing and predicting pathogen threats for specific regions. This information is needed to develop effective mitigation and rehabilitation programs. Research activities that address this need include the following:

- Integrate biological and ecological information of genetically based species groups at appropriate ecogeographic scales.
- Determine genetic and coevolutionary relationships of host-pathogen population structure and dynamics.
- Apply systematics and molecular diagnostic techniques to ensure accurate pathogen identification for integration with ecological data.
- Elucidate pathogen life cycle (e.g., growth, reproduction, survival, host infection, and spread) and epidemiology.
- Evaluate pathogen response and adaptation to hosts, host resistance factors, and environmental factors (e.g., temperature, moisture, soil properties, etc.).
- Refine remote sensing, validated by "ground truthing," to develop precise models of critical environmental factors at landscape, regional, and global scales.
- Delineate hosts and pathogen populations that behave similarly.
- Determine interrelationships of biological behavior with abiotic and biotic environments.
- Develop methods to prioritize resource allocations for managing specific pathogens.

- Provide scientific input to discussions and decisions concerning treatment measures.
- Provide science information and data for policymakers on quarantine and regulatory issues.

Impact assessment and valuation technologies are crucial for successful prioritization and management of invasive pathogens. Focused research objectives that will improve these technologies include the following:

- Understand the ecological, economic, and societal impacts of invasive pathogens.
- Increase the reliability of spatial models for projecting impacts at landscape, regional, and global scales.
- Refine procedures for cost-benefit analysis for management and remediation efforts.

Various forest disturbances exacerbate or significantly interact with invasive forest pathogens. Knowledge of these interactions is needed to strategically manage invasive pathogens. Research objectives to address this need include the following:

- Understand the influences of changes in climate, vegetation, and disturbance regimes (e.g., fire, insect outbreaks, endemic pathogens, forest fragmentation, urbanization, logging, flooding, wind, etc.) to predict distribution, ecological behavior, and effects of invasive pathogens.
- Integrate climate change models with information on the distributions of hosts and pathogens to assess their potential niches and migrational pathways.
- Understand declines and other complex diseases caused by interactions of multiple disturbances, insects, diseases, stress agents, etc., that will likely increase in changing environments at a landscape level.

Many invasive forest pathogen issues require adaptive approaches for effective management. Research outputs that address this need include the following:

- Methods that monitor and assess ecosystem function in areas affected by invasive pathogens.
- Approaches that foster resilience in forest ecosystems to minimize the impacts of future invasive pathogens.
- Techniques that enhance activities of endemic biological control agents and other natural processes to reduce the effects of invasive pathogens.

- Approaches to foster the adaptive capacity of ecosystems for recovery and renewal.

Knowledge of endemic forest pathogens and their hosts is requisite for sound assessment of threats posed by invasive pathogens. Research activities that address this need include the following:

- Coordinate with cooperators to determine baseline information on the distributions of pathogens and strains, hosts, vectors, and associated microbes, their current environments, and likely responses to changing environments; integrate these findings with FIA, FHM, the National Plant Diagnostic Network, and other continuous monitoring data.
- Develop methods that provide reliable taxonomic identification, such as molecular diagnostic methods, for application in surveys of forest pathogen distributions in the United States and other countries. These methods will allow precise documentation of potentially invasive pathogens (with the cooperation of APHIS, FHM, and international organizations).
- Develop and apply genetic and ecophysiological methods to determine the geographic distributions of host, pathogen, and microbe populations, races, and/or hybrids that display distinct ecological behavior and identify ecologically valuable host populations for conservation.
- Conduct phylogenetic and phylogeographic assessments of evolutionary and genetic relationships among hosts and pathogens worldwide.

Extensive collaborations among Forest Service deputy areas and other Federal, regional, tribal, and State governments are essential for managing invasive pathogens across large landscapes. Research input and collaboration are needed to accomplish the following:

- Implement regulations, monitoring, education, and treatments for invasive pathogens.
- Incorporate DNA-based diagnostics into strategies for the detection and monitoring of forest pathogens at ports of entry, at points of distribution, and across forest landscapes.
- Implement regional invasive species plans and coordinate these activities among agencies (e.g., via the emerging Regional Invasive Species Issue Teams).

- Establish databases and other tools to consolidate distribution data and facilitate information sharing among the natural resources research and management communities.

Available and accessible information and data on invasive pathogens are important in the development and implementation of invasive species management projects or programs. Research activities are needed to accomplish the following:

- Provide current Web sites and synthesis papers that communicate key principles and new findings to stakeholders.

- Integrate local, regional, national, and international databases that characterize forest hosts, pathogens, and associated microbes with ecological information, such as geographic distribution, climate data, digital imaging, etc., for developing predictive models and technology transfer.

- Improve the effectiveness of information delivery to those affected by invasive pathogens and develop processes to integrate communities into monitoring and research processes.

- Integrate plant pathology R&D results into invasive pathogen management activities led by FHP, FHM, NFS, States, and other national and international organizations.

Long-term research must be continued to develop management strategies for established invasive pathogens in long-lived forests, where reproductive maturity and ecological succession occur over long time periods. Research activities and research capacity are needed to accomplish the following:

- Improve the monitoring of long-term remediation effects in forest ecosystems affected by invasive pathogens.

- Apply Forest Service R&D research expertise in plant pathology, genetics, and associated disciplines to address vast expanses of forests within the United States that are not recovering from the effects of invasive plant pathogens.

- Maintain the historical role of Forest Service R&D as the primary entity responsible for conducting long-term research, maintaining data records and biological materials from established permanent field plots and supervising a network of experimental forests in the United States.

The Future of Research on Invasive Forest Pathogens

Threats posed by invasive forest pathogens are predicted to continually increase in the future due to increased human-mediated movement of pathogens and climate change. Forest Service R&D must be well positioned and well prepared to address these increased threats. Successful research programs will likely be multidisciplinary and incorporate new technologies. These programs can be directed toward five focus areas, which are discussed under five subsequent headings; the first four areas are derived from program elements of the national strategy (USDA Forest Service 2004): (1) Prediction and Prevention, (2) Early Detection and Rapid Response, (3) Management and Mitigation, (4) Restoration and Rehabilitation, and (5) Scientific Application and Communication. The following sections provide additional information on the future of research on invasive forest pathogens, based on the five focus areas.

Prediction and Prevention

Prediction and prevention is the most effective and economical strategy for managing invasive forest pathogens (Parker and Gilbert 2004). Furthermore, the value of improved methods for preventing and predicting invasive forest pathogens will increase in the future. Phylogenetic, phylogeographic, and population genetic analyses of forest hosts and pathogens on a global basis provide a novel approach for improving predictions of potentially invasive pathogens (Gilbert and Webb 2007). Such analyses determine genetic relatedness among taxa of hosts and pathogens to predict potential hosts of invasive pathogens and predict potentially invasive pathogens before they are introduced. In addition, the genetic relationships among pathogens can allow predictions of potential risks associated with the introduction of closely related species or hybridization and introgression with related endemic species (Brasier 2001, Garbelotto et al. 2004).

Climate modeling, georeferencing the distribution ranges of species across regions and landscapes, and risk analysis can be combined with genetic analyses to identify areas where invasive species may originate and areas at risk for successful establishment, invasion, and/or hybridization of invasive pathogens (Venette and Cohen 2006). Such integrated analyses can also determine the effects of climate change on the ecological behavior of invasive pathogens, and prioritize target areas for monitoring, in conjunction with risk analysis of commodity shipments from potential sources of invasive pathogens. A well-established, worldwide database, networking among involved parties (e.g., scientists, forest managers, policymakers, etc.), international collaborative research, and worldwide surveys will provide a critical basis and research infrastructure for predicting potentially invasive forest pathogens for the United States and other countries. Close cooperation with Canada and Mexico would be especially useful. Effective predictions of potentially invasive pathogens are essential to improve the efficiency of early monitoring to detect invasive pathogens for specific regions.

The development of regulations to prevent the entry and establishment of new pathogens is trending away from species-specific analyses and toward pathway approaches (Baker et al. 2005). By international consensus, these approaches still require the identification of known pests that can move in the pathway or commodity and designing mitigations to reduce the risk of specific pathogen introductions. However, broad-spectrum mitigation measures will also help prevent unknown pests that could be introduced via the identified pathways. Research is needed to identify potentially invasive pathogens and potential introduction pathways so that effective mitigation measures can be developed before they arrive. These research efforts should be conducted in collaboration with other national (e.g., APHIS, Agricultural Research Service) and international agencies that focus on high-risk, high-volume trade commodities, such as live plants, wood products and novelty items, forest seed, wood chips, and peeler cores (USDA Forest Service 2004).

Early Detection and Rapid Response

Based on past experience, improved monitoring is likely the primary activity for realistic, first detection of cryptic, invasive pathogens (Chornesky et al. 2005). Such pathogens are generally detected only after disease foci are identified in the landscape through regular monitoring efforts. Invasive forest pathogens may be eradicated or contained to a local area if detected and eradicated before widespread dispersal has occurred. For example, efforts to contain *Phytophthora ramorum* appear to have reduced its spread in Curry County, OR; however, it is yet to be determined if these efforts will remain effective over the long term (Frankel 2008). During the 20th century, significant forest pathogens remained unrecognized, misdiagnosed, or cryptic until they caused widespread damage

(e.g., oak wilt, sudden oak death, and butternut canker). In addition, invasive pathogens can arise by genetic change, evolution, or hybridization, such as what occurred in Europe with alder decline (Brasier 2001, Brasier et al. 2004).

Improved methods and tools to detect invasive forest pathogens and monitor eradication are growing needs (Chornesky et al. 2005). Emerging technologies will foster the integration of genetic and evolutionary relationships with environmental factors and thereby allow for the advanced assessment of newly discovered species. The integration of remotely sensed data, geographical information systems, soils and landform maps, climate matching software, appropriate statistical sampling models, etc., will better define target areas for detection and monitoring surveys (risk maps, e.g., Frank et al. 2008). Collaboration among detection, monitoring, survey, and inventory efforts in urban and community forests will more readily detect nonnative pathogens that may have been introduced by humans (e.g., via planting of infected nursery stock). In addition, information from controlled laboratory studies can also help confirm the potential for interspecific or intraspecific hybridization of forest pathogens and determine environmental tolerances of invasive pathogens.

DNA-based diagnostics are needed to recognize pathogen species, races, and hybrids (Crous 2005). Races or strains of pathogenic microorganisms are often not distinguishable by classical morphological and biochemical methods. In addition, traditional diagnosis of pathogens requires time-consuming host inoculation trials. Furthermore, exotic and endemic organisms can be morphologically indistinguishable but differ greatly in their ability to cause disease; even within a species, strains can behave very differently. In addition, some recent invasive forest pathogens have been shown to be hybrid species—such as the causal agents of Dutch elm disease (*Ophiostoma novo-ulmi*, Et-Touil et al. 1999), bark disease of alder (*Phytophthora alni*, Brasier et al. 2004), leaf rust of hybrid poplar (*Melampsora ×columbiana*, Newcombe et al. 2000), and a new variant of blister rust (*Cronartium ribicola × C. comandrae*, Joly et al. 2006). Representative isolates of pathogens must be archived for future study if future DNA-based technology is to be widely applicable for managing invasive forest pathogens.

Nationwide and worldwide databases are needed to compile information about potential invasions of nonnative forest pathogens (Crous 2005). Compiled and archived data on geographic distributions of existing pathogens, along with accompanying host and environmental information, are needed. ExFor (http://spfnic.fs.fed.us/exfor/index.cfm) is one example of an existing database that could be expanded to include molecular diagnostics and precise geographic locations of forest pathogens worldwide. Analysis of data from systems tracking the movement of plants (e.g., by nurseries, plant hunters and traders, etc.) will provide greater insight and predictive ability about potential introductions. For example, nursery management records, such as importation of exotic plants, location and proximity of species within nurseries, and nursery cultural practices, would be valuable in predicting potential pathogen introduction and establishment. Forest Service R&D can coordinate with APHIS and FHM to develop molecular diagnostic tools (e.g., real-time polymerase chain reaction, microarrays, etc.) to quickly detect known and potentially invasive forest pathogens in surveys of international plant shipments, ornamental and forest nurseries, and urban and periurban forests (Chornesky et al. 2005). Alternative methods of plant material movement should be developed and/or considered. For example, tissue-culture-derived plantlets and certified seed or other propagules should be developed for plants that pose high risk for moving invasive pathogens.

Management and Mitigation

Despite prediction and prevention efforts, invasive forest pathogens often escape and remain undetected until they are well established and causing noticeable damage. Thus, methods to manage invasive pathogens and mitigate their effects are needed. Ideally, management methods should foster naturalization processes that render pathogens less aggressive, hosts more resistant, and biological control agents more active over time. Other management methods are needed to enhance resiliency in forest ecosystems so that they are less affected by unforeseen threats.

Tools to manage and mitigate diseases caused by invasive pathogens must be based on sound scientific knowledge of two general processes: (1) pathogen biology, dispersal, and conditions for infection and population increase and (2) genetic and evolutionary mechanisms that control pathogenicity and aggressiveness on different hosts and in different environments. Remotely sensed imagery can be incorporated to identify distribution, effects, and spatial dynamics of invasive pathogens.

Interactions among climate, forest species compositions, fire and fuels treatments, coexisting pests, vectors, biological

control agents, and other components of a vulnerable ecosystem must be understood to determine potential establishment, reproduction, and spread of invasive pathogens (Baker and Cook 1974, Broadbent and Baker 1974, Weste and Marks 1987). Such knowledge will also aid the development of biological control and other management approaches that foster ecosystem recovery and pathogen naturalization processes.

The incorporation of environmental factors (e.g., temperature, moisture, topography, soil properties, etc.) and genetic and biological information in a spatially relational context represents the most powerful approach to understanding the ecological behavior of invasive pathogens. Integrated, multidisciplinary approaches that foster an understanding of the basic biology of each pathosystem and the interacting influences of environmental factors across spatial scales are requisite to any invasive species program. Such integrated information is used to develop stand, landscape, and regional models to aid management decisions and effectively target management efforts for maximum economic or ecological results (Chornesky et al. 2005). These approaches also will contribute to cost-benefit analyses for control actions and prioritization of areas for disease suppression activities.

Forest Service R&D has primary responsibility for long-term research studies, including maintenance of permanent plots, management of experimental forests, preservation of data archives, and maintenance of biological collections (e.g., fungal cultures, herbarium specimens, seed collections, etc.), which are critical to addressing long-term issues associated with invasive pathogens. R&D involvement in long-term invasive species management programs is needed to select and breed for pathogen-resistant trees; develop biological, chemical, and mechanical treatments to manage disease; determine interactions with fire and other disturbances; and facilitate the regeneration and recovery of ecosystems affected by invasive pathogens. Long-term research is needed to develop effective treatments and strategies in an adaptive management context. Far-sighted and diverse approaches are needed to understand and mitigate effects of invasive pathogens on diverse aspects of forest and rangeland ecosystems. Because of potential large-scale, extreme disturbances, such as drought, insect outbreak, fire, and invasive pathogens, the future regenerated forests may differ from the previous forest and may comprise multiple alternative states. The desired outcome of managing vulnerable or invaded ecosystems is adaptation or change that tolerates the disturbance but retains essential ecological functions and character. Thus, long-term research is essential to address these long-term issues posed by invasive pathogens.

Restoration and Rehabilitation

Unfortunately, numerous examples have occurred where invasive forest pathogens became established, proliferated, and caused widespread damage to forest ecosystems across regions of the United States and elsewhere. In most of these cases, research efforts were abandoned or severely curtailed before affected ecosystems were on track toward recovery (Tainter 2003). Similar to many medical and environmental issues, many invasive forest pathogens cause long-term damage to forest ecosystems that can only be effectively addressed through long-term restoration efforts supported with research and monitoring. Prioritization is needed to determine which damaged ecosystems justify the expenditure of funds for restoration and rehabilitation.

Because distinct pathogen populations respond differently to environmental influences, it is important to consider host and pathogen population structures when assessing threats posed by invasive pathogens before implementing restoration strategies (Burdon et al. 2006). Basic information needs for mitigating invasive pathogen effects include identification of sources of host resistance (e.g., through screening and selection programs); understanding of resistance mechanisms; increasing host resistance through breeding or regeneration programs; the development of rational deployment strategies for resistant plant materials; and identification of threatened, ecologically valuable host populations for conservation. Landscape-level, ecosystem-based research is also needed to develop techniques for restoring ecosystem function and structure, enhancing natural biological controls, fostering natural recovery and stabilization, and improving the resiliency of ecosystems to unexpected future threats (Drever et al. 2006).

Scientific Application and Communication

The primary clients for new information and tools for managing invasive pathogens include natural resource managers, regulatory officials, policymakers, tribal governments, private industry entities, nongovernmental organizations, research scientists, educators, and the public at large (USDA Forest Service 2007). Thus, current and future issues regarding invasive forest pathogens require effective communication through diverse venues (e.g., Web-based tools, workshops, online training, scientific/technical publications, etc.). It is essential that research is directed toward obtaining actionable knowledge regarding invasive pathogen management, and research scientists should be active participants in the communication processes that make research results available and applicable to the clients. Participation in international working groups, such as the International Union of Forest Research Organizations Working Party 7.03.12 (alien invasive species and international trade), is also essential for building international collaborations and exchanging information about invasive forest pathogens.

Critical information regarding pathogen genetic identification, biology, hosts, geographic and ecological distribution (e.g., climate data, slope, aspect, habitat type, soil properties, and other environmental factors), available research literature, etc., should be readily accessible from online databases. Other Internet technology is needed to address specific invasive pathogen issues and connect professionals with different areas of expertise. To maintain quality, ensure accountability, and promote the appropriate use of information and models, the communication process needs to include standard features of the scientific process, such as replication and review. The hypertext capability of digital media should be used to augment information with metadata on sources and other supplemental or supporting documents. Given that uncertainty and complexity are inherent, it is imperative that research maintains credibility and science is responsibly applied.

Raising public awareness of invasive forest pathogen issues is also critical. Social science research is needed to develop effective methods that encourage the adoption of practices across ownerships. Within landscapes, this research is needed to avoid the spread of invasive pathogens and gain an understanding and acceptance of management practices that may be considered as an interference to the public's daily life. The Forest Service also must strive to develop more diverse program delivery methods (e.g., social networking Web sites, podcasts, streaming video, cell phone-mediated information retrieval, television, radio, magazines, partnering with commercial franchise marketing, multilingual communication, etc.) and other integrated tools developed for diverse audiences.

Acknowledgments

The authors thank Phil Cannon (Region 5, California), Kerry Britton (R&D-WO), and three anonymous reviewers for comments on an earlier version of this manuscript.

Literature Cited

Anagnostakis, S.L. 1987. Chestnut blight: the classical problem of an introduced pathogen. Mycologia. 79: 23–37.

Baker, K.B.; Cook, R.J. 1974. Biological control of plant pathogens. San Francisco: W.H. Freeman & Co. 433 p.

Baker, R.; Cannon, R.; Bartlett, P.; Barker, I. 2005. Novel strategies for assessing and managing the risks posed by invasive alien species to global crop production and biodiversity. Annals of Applied Biology. 146: 177–191.

Brasier, C.M. 2001. Rapid evolution of introduced plant pathogens via interspecific hybridization. BioScience. 51: 123–133.

Brasier, C.M.; Buck, K.W. 2001. Rapid evolutionary changes in a globally invading fungal pathogen (Dutch elm disease). Biological Invasions. 3: 223–233.

Brasier, C.M.; Kirk, S.A.; Delcan, J., et al. 2004. *Phytophthora alni* sp. nov. and its variants: designation of emerging heteroploid hybrid pathogens spreading on *Alnus* trees. Mycological Research. 108: 1172–1184.

Broadbent, P.; Baker, K.F. 1974. Behavior of *Phytophthora cinnamomi* in soils suppressive and conducive to root rot. Australian Journal of Agricultural Research. 25: 121–137.

Burdon, J.J.; Thrall, P.H.; Ericson, L. 2006. The current and future dynamics of disease in plant communities. Annual Review of Phytopathology. 44: 19–39.

Chornesky, E.A.; Bartuska, A.M.; Aplet, G.H., et al. 2005. Science priorities for reducing the threat to invasive species to sustainable forestry. BioScience. 55: 335–348.

Crous, P.W. 2005. Impact of molecular phylogenetics on the taxonomy and diagnostics of fungi. OEPP/EPPO Bulletin. 35: 47–51.

Daughtrey, M.L.; Hibben, C.R.; Britton, K.O., et al. 1996. Dogwood anthracnose: understanding a disease new to North America. Plant Disease. 80: 349–358.

Drever, C.R.; Peterson, G.; Messier, C., et al. 2006. Can forest management based on natural disturbances maintain ecological resilience? Canadian Journal of Forest Research. 36: 2285–2299.

Et-Touil, A.; Brasier, C.M.; Bernier, L. 1999. Localization of a pathogenicity gene in *Ophiostoma novo-ulmi* and evidence that it may be introgressed from *O. ulmi*. Molecular Plant-Microbe Interactions. 12: 6–15.

Fraedrich, S.W.; Harrington, T.C.; Rabaglia, R.J., et al. 2008. A fungal symbiont of the redbay ambrosia beetle causes a lethal wilt in redbay and other Lauraceae in the southeastern United States. Plant Disease. 92: 215–224.

Frank, K.L.; Geils, B.W.; Kalkstein, L.S.; Thistle, H.W., Jr. 2008. Synoptic climatology of the long distance dispersal of white pine blister rust. II. Combination of surface and upper-level conditions. International Journal of Biometeorology. 52: 653–666.

Frankel, S.J. 2008. Sudden oak death and *Phytophthora ramorum* in the USA: a management challenge. Australasian Plant Pathology. 37: 19–25.

Furnier, G.R.; Stolz, A.M.; Mustaphi, R.M.; Ostry, M.E. 1999. Genetic evidence that butternut canker was recently introduced into North America. Canadian Journal of Botany. 77: 783–785.

Garbelotto, M.; Gonthier, P.; Linzer, R., et al. 2004. A shift in nuclear state as the result of natural interspecific hybridization between two North American taxa of the basidiomycete *Heterobasidion*. Fungal Genetics and Biology. 41: 1046–1051.

Geils, B.W.; Hummer, K.E.; Hunt, R.S. 2010. White pines, Ribes and white pine blister rust: a review and synthesis. Forest Pathology. 40: in press.

Gilbert, G.S.; Webb, C.O. 2007. Phylogenetic signal in plant pathogen-host range. Proceedings of the National Academy of Science (USA). 104: 4979–4983.

Gordon, T.R.; Storer, A.J.; Wood, D.L. 2001. The pitch canker epidemic in California. Plant Disease. 85: 1128–1139.

Hamelin, R.C.; Joly, D.; Langor, D. 2005. A hybrid between white pine blister rust and Comandra blister rust on limber pine. Phytopathology. 95 (supplement): S39–40.

Hamelin, R.C.; Lecours, N.; Laflamme, G. 1998. Molecular evidence of distinct introductions of the European race of *Gremmeniella abietina* into North America. Phytopathology. 88: 582–588.

Joly, D.L.; Langor, D.W.; Hamelin, R.C. 2006. Molecular and morphological evidence for interspecific hybridization between *Cronartium ribicola* and *C. comandrae* on *Pinus flexilis* in southwestern Alberta. Plant Disease. 90: 1,552.

Lovett, G.M.; Canham, C.D.; Arthur, M.A., et al. 2006. Forest ecosystem responses to exotic pests and pathogens in eastern North America. BioScience. 56: 395–403.

Maloy, O.C. 1997. White pine blister rust control in North America: a case history. Annual Review of Phytopathology. 35: 87–109.

Newcombe, G.; Stirling, B.; McDonald, S.; Bradshaw, H.D., Jr. 2000. *Melampsora × columbiana*, a natural hybrid of *M. medusae* and *M. occidentalis*. Mycological Research. 104: 261–274.

Otrosina, W.J. 2005. Exotic ecosystems: where root disease is not a beneficial component of temperate conifer forests. In: Lundquist, J.E.; Hamelin, R.C., eds. Forest pathology—from genes to landscapes. St. Paul, MN: APS Press: 121–126.

Palm, M.E. 2001. Systematics and the impact of invasive fungi on agriculture in the United States. Bioscience. 51: 141–147.

Parker, I.M.; Gilbert, G.S. 2004. The evolutionary ecology of novel plant-pathogen interactions. Annual Review of Ecology Evolution and Systematics. 35: 675–700.

Rizzo, D.M.; Garbelotto, M.; Hansen, E.M. 2005. *Phytophthora ramorum*: integrative research and management of an emerging pathogen in California and Oregon forests. Annual Review of Phytopathology. 43: 309–335.

Tainter, F.H. 2003. Perspectives and challenges from the 20th century. Phytopathology. 93: 1056–1061.

U.S. Department of Agriculture (USDA) Forest Service. 2004. National strategy and implementation plan for invasive species management. Washington, DC: U.S. Department of Agriculture, Forest Service. 18 p. Available at http://www.fs.fed.us/invasivespecies/documents/Final_National_Strategy_100804.pdf.

USDA Forest Service. 2007. Cohesive approach for invasive species management in the Northeastern U.S. Newtown Square, PA: U.S. Department of Agriculture, Forest Service, Northeastern Area, Northern Research Station, Region 9. 11 p.

Venette, R.C.; Cohen, S.D. 2006. Potential climatic suitability for establishment of *Phytophthora ramorum* within the contiguous United States. Forest Ecology and Management. 231: 18–26.

Weste, G.; Marks, G.C. 1987. The biology of *Phytophthora cinnamomi* in Australasian forests. Annual Review of Phytopathology. 25: 207–229.

The Role of the Forest Service in Nonnative Invasive Plant Research

Carolyn Hull Sieg[1], Julie S. Denslow[2], Cynthia D. Huebner[3], and James H. Miller[4]

Abstract

In many of our Nation's wildlands, invasive nonnative plants contribute to the endangerment of native species and lead to other severe ecological and financial consequences. Projected trends of increasing human populations and associated development and globalization will contribute to increases in the already high rates of introductions of nonnative plant species. Changes in climate are likely to alter species distributions, favoring the expansion of some nonnative species and contributing to the imperilment of additional native species. Declining oil supplies may also place pressure on wildlands for the production of sustainable supplies of small-diameter trees or other nonwoody biofuels. Given these trends, Forest Service Research and Development needs to be strategic in addressing invasive species issues in public and private forests and rangelands. We urgently need to prioritize both known and potential future invasive species and determine which ecosystems are most vulnerable to invasion. Quantitative risk analyses, assessment of critical pathways, plus data on effects of both the invaders and control methods on native biodiversity will aid in this prioritization process. Such lists will inform decisionmaking on potential preventative measures to keep potentially invasive plants out and also as a guide regarding which species to attempt to control and where to control them. Multidisciplinary research teams and quantitative monitoring protocols will facilitate the development of tools that both measure and minimize effects associated with invasive species and account for the stage of invasion. These tools will also need to address multiple stressors, including natural disturbances, current management practices such as livestock grazing and timber harvesting and thinning, and human-induced disturbances, such as exotic insect forest infestations and global climate change. Such knowledge will improve our ability to manage our forests and rangelands as ecosystems that are more resilient to future invasions and increase our success in restoring degraded systems.

Introduction

The introduction and spread of nonnative plant species are natural colonization and migration processes that occurred prior to human evolution, and such invasions have been documented by several early researchers (Darwin 1859, Elton 1958). However, recorded invasions of exotic plants and other exotic species since 1800 have increased at an accelerated rate, presumably due to increased intercontinental mobility (Liebhold et al. 1995). In the United States, plant introductions are currently allowed without prior risk assessments. While many nonnative plant species provide food and fiber, adorn our civilization, and facilitate habitat restoration and land management, some will spread widely and possibly alter ecosystem structure and processes in undesirable ways. These invasive species incur high costs in altered ecosystem services and in investments in their management and control. Invasive exotic plants constitute 8 to 47 percent of the total flora of most States in the United States (Rejmanek and Randall 1994). There are approximately 4,500 exotic species in the United States that have established naturalized populations and at least 15 percent of these cause severe harm (U.S. Congress, Office of Technology Assessment 1993).

Examples of negative ecosystem effects caused by invasive plants include alteration of food webs (Bailey et al. 2001, Kourtev et al. 1999), degradation of wildlife habitat (Schmidt and Whelan 1999), changes in fire (Brooks et al. 2004) and hydrological regimes (Gordon 1998), increases in erosion rates

---

[1] Research Plant Ecologist, Forest Service, Rocky Mountain Research Station, 2500 South Pine Knoll Drive, Flagstaff, AZ 86001.

[2] Scientist Emerita, Forest Service, Institute of Pacific Island Forestry, 60 Nowelo Street, Hilo, HI 96720.

[3] Research Botanist/Ecologist, Forest Service, Northern Research Station, 180 Canfield Street, Morgantown, WV 26505–3180.

[4] Research Ecologist, Forest Service, Southern Research Station, 520 Devall Place, Auburn, AL 36849.

(Shafroth et al. 2002), and modifications of nutrient cycling (Ehrenfeld et al. 2001, Sperry et al. 2006). Nonnative plant species can also reduce biological and genetic diversity by hybridizing with native plants (White and Bowden 1947). It is estimated that the United States spends approximately $145 million annually in its attempt to control nonnative invasive plants in natural areas (Pimentel et al. 2000).

Forest Service Research and Development (R&D) plays an important role in meeting research needs involving invasive nonnative species in the Nation's forests and rangelands. Nevertheless, scarce resources mandate that the Forest Service focus its efforts strategically. This document outlines priority Forest Service R&D needs relating to invasive plant species threatening these wildlands.

Effective prioritization of research resources entails recognition of the wide variation in invasive effects, ecosystem vulnerabilities, and ecological and economic consequences of the spread of a nonnative species. In many second-growth, closed-canopy forests, some populations of nonnative species are small or confined to forest edges or canopy gaps; such species may be controlled. Without such efforts, areas only lightly invaded may become more severely affected in response to forest management, new disturbances, and/or a proliferation of invasive species (especially those that are shade-tolerant) propagules, threatening ecosystem structure and processes and the economic, environmental, and societal benefits derived from them. In other cases, nonnative species posing considerable risk to ecosystem integrity have spread widely and reached high densities, thus altering forest management priorities, and requiring long-term investments to mitigate effects. The Forest Service, with guidance from local, State, and Federal governments, must balance the relative value of wildland ecosystems, the potential magnitude of invasive effects, and the costs to society in its decisions to invest research resources into understanding these processes and developing tools to address them.

Multiple pathways for introductions and increasingly fragmented landscapes mean that Forest Service lands cannot be effectively managed without consideration of the landscapes in which they occur. Landscape and regional perspectives on invasion processes encourage cooperation with adjacent landowners, municipalities, and other agencies to address invasive species issues on national forest lands.

Strengths of Forest Service Research and Development

The Forest Service role results from its unique ability to make a difference nationally. The Forest Service has broad existing authorities and responsibilities, assigned to the Chief of the Forest Service; research expertise in land management, forestry, entomology, pathology, botany, ecology, and numerous other specialties; presence across the country and around the world; and relationships with every State and territorial agency with responsibility for invasive species. Few other agencies, universities, and organizations have such a long-term perspective on land management and research directions or such valuable assets for attracting partners (including long-term data sets, experimental forests, research natural areas, and quarantine facilities).

Forest Service researchers have a long history of collaboration with other agencies and research partners, facilitating effective management of invasive plants and identification of priority research issues and needs addressed at appropriate regional and landscape scales. Close associations with partners in universities, industry, and other Federal agencies facilitate collaborations to supplement Forest Service expertise. The geographic distribution of Forest Service research stations and broad land base that encompasses a wide variety of forests, woodlands, shrublands, and grasslands provide excellent opportunities to test research hypotheses and models at multiple geographic and temporal scales. The Forest Inventory and Analysis system within Forest Service R&D houses long-term national, regional, and local data sets that have recently been extended to collect data on some nonnative plant species in monitored plots. The broad research expertise among Forest Service scientists facilitates a multidisciplinary approach to the study of invasive species, development of tools for their management, and protocols for forest and rangeland restoration.

Key Future Issues

The 2000 Renewable Resources Planning Act (RPA) assessment (RPA 2000) projects growing U.S. populations (50-percent increase by 2050), especially in the southern and western regions. Population increases are expected to increase demand for forest services (especially recreational uses) and increase the conversion of forests and rangelands to developed use, resulting in further subdivision and fragmentation. The forest land base is expected to remain relatively stable. The 2000 RPA projects aging forests in many parts of the country, changing their vulnerability to invasive species, insect and disease outbreaks, and fire. Projected changes in climate and atmospheric inputs such as nitrogen will affect species distributions and nutrient cycles and the duration, frequency, and intensity of forest disturbances such as fire, insects, diseases, drought, and storms. These scenarios, in combination with growing global trade, suggest that threats and effects of invasive plant species are likely to increase in the next 20 to 50 years, challenging the Forest Service research community to address landscape, regional, and national issues of invasive species management and mitigation.

Our ability to manage nonnative plant species invasions is linked with past, current, and future human values and associated land management activities as well as with sometimes conflicting priorities over the use of wildlands. For example, larger, more frequent, and more intense wildfires in dense and infrequently burned forests may enhance the spread of some disturbance-dependent invasive species. The spread of invasive nonnative plants associated with recreation, roads, habitat fragmentation, grazing, harvesting, tree thinning, prescribed fires, and fuel reduction activities may enhance populations of invasive species. The cultivation of potentially invasive plant species as biofuels could facilitate the spread of invasive species into the Nation's forests and natural areas. Furthermore, the use of forest lands to provide sustainable supplies of small-diameter trees to support biomass power could have the unintended consequence of introducing invasive plants into managed forests at unprecedented rates.

Priority setting for invasive species research should reflect societal values with respect to species, ecosystems, and ecological services. Is our highest priority to preserve ecosystem processes that are dynamic or to preserve a static ecosystem structure (Botkin 2001)? How do we address the needs of imperiled species against a landscape undergoing dramatic changes? How do we address changing priorities for ecosystem services given dramatic shifts in human population sizes, ethnic diversity, and age structure? How do we focus research on target species when confronted with rising rates of invasion?

Prevention and Prediction

Preventing the widespread establishment of nonnative species may be more cost effective than attempting to control full-blown infestations, which may not be economically feasible. Quantitative analyses are needed to better understand the distribution and abundance patterns of nonnative species populations, pathways of introduction, and habitats most at risk. The development of predictive models that identify areas likely to be negatively affected by nonnative species and accounting for sampling effort, climate, physiography, human population density, and other variables that reflect land use intensity is an important step in developing a national invasive species strategy. The concurrent development of improved pathway analytical methods that identify probable entry points and means and modes of introduction, establishment, and spread will support the development of stronger prevention programs. Research that addresses the social aspects of nonnative plant use will enhance our ability to institute prevention guidelines suitable to different cultural conditions.

Our poor understanding of factors that make plant communities susceptible to invasion limits our ability to provide management guidelines for preventing the introduction and spread of nonnative plant species. Ecosystem attributes, disturbances, and plant characteristics can all influence invasion rates. A better understanding of the role of stand structural attributes and altered disturbance regimes is needed, as is plant-focused research. High rates of pollen and seed rain from invasive plants and the long-term viability of seed in the soil may overwhelm the biotic resistance of even the most nonfragmented and native plant-dominated ecosystems. We need to better understand the roles of propagule pressure and the numbers, sizes, and distributions of invasive plant populations to better predict the dynamics of spread. Pathway analyses and models of spread should incorporate estimates of dispersal distances and predictions of safe establishment sites, both of which require a detailed understanding of each plant's reproductive strategies and physiology at various stages of development.

For both large-scale models of invasive species patterns and local-scale studies of the role of stand-level attributes, high-quality data sets are needed. Forest Service researchers will need to collaborate with universities and States to ensure that data on nonnative species are collected, verified, and made available in Web-accessible databases. Ready access to taxonomic experts is critical for detecting and responding to populations of invasive species when they are small and manageable. Likewise, investments in forest monitoring plots, such as the Forest Inventory and Analysis system, may provide information on invasions relative to stand structural attributes and assist in predicting invasion trajectories and potential effects of invasions on community productivity and diversity.

Early Detection and Rapid Response

There is a need for a centralized, rapid, effective procedure for identifying potential new plant invaders. A user-friendly database of information on potential invaders and management methods would facilitate early response. Effective survey strategies and techniques are needed for detecting very small, isolated populations of newly established nonnative species and for predicting invasible habitats, perhaps based on habitat risk and vector assessments. Research is needed to determine the best tools for land-based, regional surveys of nonnative invasive species that are rapid, quantifiable, and repeatable.

We need to expand our proactive research role that will promote early and rapid management of plant invasions by Federal, State, and local entities when such invasions are small and controllable. Risk assessments at this stage should be based on plant population ecology and metapopulation models, especially dispersal rates, pathways, and distances. There is a need to develop tools to identify potentially invasive species, prioritize species for management attention, identify priority areas for treatment, and more successfully eradicate high-priority invaders with minimal nontarget effects. Spread models and forest growth and regeneration models that use various scenarios to predict species compositional changes (and associated economic losses in property values) over time and under different levels of control and management (including the do-nothing alternative) will provide scientific- and economic-based incentives to respond early and rapidly. There is a need for social science research focused on understanding the human behavior of those responding to the invasions, because the success of any new rapid response tool is dependent on the cooperation of a number of different people, including scientists, land managers, and amateur botanists, as well as public response.

Management and Mitigation

Rising rates of introduction coupled with the spread of increasing numbers of invasive plant species on forests and rangelands will challenge land managers to determine priorities for control and mitigation. Research quantifying ecological and socioeconomic effects of nonnative invasive plant species is critically needed to aid decisionmaking, focus management efforts, and develop a better understanding of the behavior of different species and ecosystems under different environmental circumstances. What are the likely long-term consequences of no management for a newly invasive plant species? Forest Service research into tool development should seek to maximize effects on target organisms and to minimize nontarget effects. In most cases, knowledge of the basic biology, population genetics, and population ecology of high-priority target species will be necessary to design protocols to mitigate their effects. Managers may also seek lessons learned from similar species and ecosystems to develop general protocols. Genetic changes in populations due to selection, hybridization, and the introduction of new genotypes may increase invasiveness; there is a need to understand the contribution of genetic variation in driving plant invasions.

Invasive plant managers will be able to take advantage of new tools, including remote sensing, genetic evaluation, landscape analysis, epidemiological modeling, and statistics. Integrated pest management using mechanical and chemical treatments, biological control agents, and vegetation management via prescribed fire and grazing will continue to play an important role in the mitigation of invasive plant effects. Research into the application, integration, and effects of a broad spectrum of tools is needed to improve efficacy, expand our ability to treat different ecosystem types, reduce undesirable effects, and address emerging invasive species. There is also a need for greater emphasis on the development of cost-effective and sensitive quantitative monitoring protocols to better assess the effectiveness of various control strategies.

Restoration and Rehabilitation

Mitigation of invasive effects and increasing resistance to future invasions are accomplished through restoration and rehabilitation activities. We need research to develop vegetation management protocols to restore ecosystem processes and reduce the effects of invasives, also recognizing those situations where restoration is neither feasible nor desirable (Lugo and Helmer 2004). There is a need to develop ecologically sound restoration methods that consider the contributions of genetics, population and community structure, and ecosystem processes to invasion resistance. Disturbance, high-propagule pressures, legacy effects such as seed banks, and changes in disturbance regimes and nutrient supply may impede long-term restoration success.

Land managers lack many basic tools for reducing the effects of invasive species on severely infested lands. Forest Service research can develop propagation techniques and reestablishment tools (Mahalovich and McArthur 2004, Monsen et al. 2004), address the consequences of using nonnative plants in rehabilitation, reduce seed contamination, and guide managers in decisionmaking. Forest Service researchers can develop guidelines on appropriate species and genotypes for rehabilitation projects.

Application and Communication

Effective communication and application of invasive species research will benefit from associated sociological research. There is a need to know how best to educate forest workers, landowners, agencies, public land users, nurseries, and highway departments on the importance of preventing the introduction of nonnative species, eradicating priority species, and restoring areas degraded by nonnative species invasions. Invasive plant management guidelines that have been carefully researched should be incorporated into existing forest management models and tools.

Web-based databases and networks on invasives are maintained by the U.S. Department of Agriculture (USDA) Natural Resources Conservation Service's PLANTS database (http://plants.usda.gov), National Invasive Species Information Center's Web site (http://www.invasivespeciesinfo.gov), National Park Service's WeedUS database (http://www.nps.gov/plants/alien/), regional and State Exotic Pest Plant Councils, Invasive Plant Councils, national heritage programs, and conservation data centers, e.g., Natureserve's http://www.natureserve.org/aboutUs/network.jsp. Forest Service research should contribute

to these resources by providing science findings, syntheses, and guides. Nonetheless, traditional scientific publications will continue to provide the foundation of peer-reviewed knowledge about invasive plant species.

Timely and effective application of Forest Service research will require increased science delivery, communication, and collaboration with managers and regulatory agencies, including the USDA Animal and Plant Health Inspection Service and State agricultural and natural resource departments. The development of multiagency networks may enhance timely science application.

Future Top Priorities

- Quantitative risk analyses are needed to identify species that should not be brought into the United States or sold in nurseries and to prioritize research on individual species or species groups.

- Pathway analysis is needed to identify key pathways for species' introductions, vectors of species spread, probable points of entry for surveillance for early detection, education programs, and management planning.

- A better understanding of the shared characteristics, behaviors, and environmental thresholds of successful plant invaders is needed.

- Researchers and managers need ready access to taxonomic expertise for reliable early detection and risk assessment.

- The timely application of research results requires active interaction of researchers and managers as well as other technology transfer vehicles such as workshops for landowners and up-to-date information on Web sites and other user-friendly formats.

- There is a critical need to better understand the effects of invasive plant species on native biodiversity and on ecosystem services and to develop protocols that assess and reduce the effects of multiple stressors such as disturbances, climate change, and invasive species on rare and endangered species.

- Research is needed to aid in the recognition of habitats most vulnerable to invasions and to potential loss of biodiversity and alteration of ecosystem services.

- Research is needed to develop management and restoration strategies for high-priority species and high-priority habitats.
- We need a better understanding of the effects of different tools used to manage invasive species, their nontarget effects, and whether benefits are real.
- The development of "virtual" research teams that cut across regions and invasive species taxa should be used to better leverage Forest Service expertise in invasive species research. For example, Forest Service biological control efforts could be more synergistic, involving multiple areas of expertise and including Forest Service botanists, entomologists, ecologists, pathologists, sociologists, and people with expertise in monitoring plant population trends and nontarget effects.

Acknowledgments

The authors appreciate the contributions of Tim Harrington, Catherine Parks, Richard Cronn, Marty Vavra, and Becky Kerns (Forest Service Pacific Northwest Station); Dean Pearson, Jim Fowler, Durant McArthur, Steve Sutherland, and George Markin (Forest Service Rocky Mountain Research Station); and Ron Thill (Forest Service Southern Research Station). In addition, Guy McPherson (University of Arizona), Nancy Loewenstein (Auburn University), David Moorhead (University of Georgia), Keith Clay (Indiana University), John Byrd (Mississippi State University), Jill Swearingen (National Park Service), Brian McCarthy (University of Ohio), Lewis Ziska (Agricultural Research Service), plus Janette Kaiser, Robert Mangold, Mike Ielmini, and an anonymous reviewer (Forest Service) provided comments that greatly improved an earlier draft.

Literature Cited

Bailey, J.K.; Schweitzer, J.A.; Whitham, T.G. 2001. Salt cedar negatively affects biodiversity of aquatic macroinvertebrates. Wetlands. 21(3): 442–447.

Botkin, D.B. 2001. The naturalness of biological invasions. Western North American Naturalist. 61: 261–266.

Brooks, M.L.; D'Antonio, C.M.; Richardson, D.M., et al. 2004. Effects of invasive alien plants on fire regimes. BioScience. 54: 677–688.

Darwin, C. 1859. On the origin of species by means of natural selection, or the preservation of favoured races in the struggle for life. London, England: John Murray.

Ehrenfeld, J.G.; Kourtev, P.; Huang, W. 2001. Changes in soil functions following invasions of exotic understory plants in deciduous forests. Ecological Applications. 11(5): 1287–1300.

Elton, C.S. 1958. The ecology of invasions by animals and plants. London, England: Chapman and Hall.

Gordon, D.R. 1998. Effects of invasive, non-indigenous plant species on ecosystem processes: lessons from Florida. Ecological Applications. 8(4): 975–989.

Kourtev, P.S.; Huang, W.Z.; Ehrenfeld, J.G. 1999. Differences in earthworm densities and nitrogen dynamics in soils under exotic and native plant species. Biological Invasions. 1: 237–245.

Liebhold, A.M.; MacDonald, W.L.; Bergdahl, D.; Mastro, V.C. 1995. Invasion by exotic forest pests: a threat to forest ecosystems. Forest Science Monograph. 30: 1–49.

Lugo, A.E.; Helmer, E. 2004. Emerging forests on abandoned land: Puerto Rico's new forests. Forest Ecology and Management. 190: 145–161.

Mahalovich, M.F.; McArthur, E.D. 2004. Sagebrush (*Artemisia* spp.) seed and plant transfer guidelines. Native Plants. 5: 141–148.

Monsen, S.B.; Stevens, R.; Shaw, N.L., comps. 2004. Restoring western ranges and wildlands. Gen. Tech. Rep. RMRS-GTR-136. Fort Collins, CO: U.S. Department of Agriculture, Forest Service, Rocky Mountain Research Station. 3 vol. 884 p.

Pimentel, D.; Lach, L.; Zuniga, R.; Morrison, D. 2000. Environmental and economic costs of nonindigenous species in the United States. BioScience. 50: 53–65.

Rejmanek, M.; Randall, J. 1994. Invasive alien plants in California: 1993 summary and comparison with other areas of North America. Madrono. 41(3): 161–177.

Renewable Resources Planning Act (RPA). 2000. http://www.fs.fed.us/pl/rpa/rpaasses.pdf. (May 17, 2007).

Schmidt, K.A.; Whelan, C.J. 1999. Effects of exotic *Lonicera* and *Rhamnus* on songbird nest predation. Conservation Biology. 13(6): 1502–1506.

Shafroth, P.B.; Stromberg, J.C.; Patten, D.T. 2002. Riparian vegetation response to altered disturbance and stress regimes. Ecological Applications. 12(1): 107–123.

Sperry, L.J.; Belnap, J.; Evans, R.D. 2006. *Bromus tectorum* invasion alters nitrogen dynamics in an undisturbed arid grassland ecosystem. Ecology. 87(3): 603–615.

U.S. Congress, Office of Technology Assessment. 1993 (September). Harmful non-indigenous species in the United States. OTA-F-565. Washington, DC: U.S. Government Printing Office.

White, O.E.; Bowden, W.M. 1947. Oriental and American bittersweet hybrids. Journal of Heredity. 38(4): 125–127.

Terrestrial Animals as Invasive Species and as Species at Risk From Invasions

Deborah M. Finch[1], Dean Pearson[2], Joseph Wunderle[3], and Wayne Arendt[3]

Abstract

Including terrestrial animal species in the invasive species strategy plan is an important step in invasive species management. Invasions by nonindigenous species threaten nearly 50 percent of imperiled native species in the United States and are the Nation's second leading cause of species endangerment. Invasion and conversion of native habitats by exotic species can have detrimental effects on animal species by reducing habitat quality through changes in habitat structure, shelter, food availability, and community interactions. Managers need information about invasive animals and native animal responses to invasions to prepare management plans. Regulatory laws such as the Endangered Species Act require that potential effects on animal populations be evaluated before taking action to alter or restore habitats. "Injurious" invasive animal species must also be regulated under the Lacey Act.

The Forest Service is well positioned to address problems caused by invasive animals and mitigate effects of invasive exotic species on native animals. National forests and grasslands provide diverse habitats for numerous wildlife species. The Forest Service has scientists, ongoing studies, networks, partnerships and experimental forests and ranges focused on understanding problems linking animal species, invaders, and habitat changes.

Many of our customers are currently based in rural areas, but customer demand can be expected to shift over the next 50 years to urban communities as rural areas become urbanized. Our range of customers will expand worldwide as invasive species problems become increasingly global. Preventing global homogenization, or the ecological replacement of native species with widespread exotics, will require global communication. It is imperative that the Forest Service take an active international role in communicating solutions about this topic to a global community.

Key future issues for terrestrial animals include the following:

- Protecting wildlife from endangerment by invasions.
- Rehabilitating invaded riparian habitats and conserving riparian-obligate species.
- Managing damage to wildlife habitats from introduced insect species.
- Restoring and rehabilitating invaded wildlife habitats prone to fire outbreaks.
- Conserving animal species affected by habitat conversion and fragmentation.
- Managing wildlife habitats in relation to invasions propagated by climate change.
- Managing urban-wildland areas invaded by high numbers of nonnative species.
- Mitigating and managing the effects of nonnative diseases and viruses spread by animals.
- Detecting and eradicating invasive species in areas of high wildlife diversity.
- Understanding patterns and movements of invasive organisms across landscapes.

The top five priorities for terrestrial animals are as follows:

1. Develop knowledge and tools to manage and mitigate the effects of invasions facilitated by globalization, urbanization, and climate change on distributions and abundances of native animal species.

2. Develop knowledge and tools to improve the effectiveness of invasive species management and minimize management side effects.

[1] Research Wildlife Biologist, Forest Service, Rocky Mountain Research Station, 333 Broadway SE., Suite 115, Albuquerque, NM 87102.

[2] Research Ecologist, Forest Service, Rocky Mountain Research Station, 790 East Beckwith Avenue, Missoula, MT 59801.

[3] Research Wildlife Biologist, Forest Service, International Institute of Tropical Forestry, P.O. Box B, Palmer, PR 00721.

3. Develop knowledge and tools for managing invaded woodland, desert, steppe, and grassland ecosystems to conserve native species.

4. Develop knowledge and tools to manage and rehabilitate invaded riparian ecosystems, recover native species, and conserve biological diversity.

5. Develop knowledge and tools to mitigate invader effects on native species and habitats in tropical island ecosystems.

Why Terrestrial Animals Need To Be Included in the Invasive Species Strategy Plan

Invasions by nonindigenous species threaten nearly 50 percent of imperiled native species in the United States and are the Nation's second leading cause of species endangerment after habitat destruction and degradation (Wilcove et al. 2000, Wilson 2002). Population and ecosystem effects by invasive species include disease, predation, competition, parasitism, hybridization, alteration of disturbance regimes, alteration of nutrient cycles, and alteration of hydrologic cycles, all of which can affect terrestrial animal species (Mack et al. 2000).

Native animals are adapted to habitats composed of native plants. Invasion and conversion of native habitats by nonnative plant species can have detrimental effects on animal species by altering habitat quality through alterations in habitat structure, hiding and shading cover, food abundance, arthropod emergence cycles, nesting and denning substrates, animal species composition, predation rates, parasitism rates, and competitive interactions. Animal population responses to invaded or converted habitats can include species population declines, reduced species productivity, behavioral changes, disruptions in breeding cycles, emigration, and endangerment of populations or species.

We need to understand the relationships between changes in plant communities caused by invasions and animal populations to develop habitat restoration and management plans. Regulatory laws such as the Endangered Species Act, Migratory Bird Research and Management Act, and National Environmental Policy Act require that effects on animal populations be evaluated before altering or restoring habitats through management actions. Thus, managers need research information about the status of animal populations before and during restoration and management of invaded habitats. In addition, many terrestrial animal species, such as the brushtail possum (*Trichosurus vulpeculia*) and the brown tree snake (*Boiga irregularis*), are known to be highly invasive once introduced and are identified as "injurious" by the U.S. Fish and Wildlife Service, requiring regulation under the Lacey Act.

Mobile animal species can carry and spread the etiological agents of emerging diseases such as West Nile virus and avian influenza virus. Knowledge of animal behavior, infection rates, and patterns of movement and migration of infected animals is needed to reduce animal mortality rates and prevent spread of diseases to humans. Mobile animal species also disperse seeds, thus potentially furthering the spread of nonnative plant species. Development of knowledge of animal movement and seed dispersal patterns may help predict patterns of species invasion and prevent further spread.

Some animal species may have adapted over time to the presence of nonnative species, and restoring habitats to their original condition can have negative consequences to the native animal species, particularly if it is endangered. We need to better understand the potential negative consequences of restoration to mitigate such effects.

What Is the Unique Role for the Forest Service Today?

- The Forest Service manages a wide network of national forests and grasslands for natural resources, ecosystem services, and recreation. These public lands provide diverse habitats for numerous animal species. The Forest Service must safeguard its wildlife resources as it has been entrusted to do by the public. Nonnative species are identified as the second greatest cause of species imperilment (Wilcove et al. 2000). The Forest Service is mandated by regulatory law and by forest plans to manage wildlife and their habitats, prevent and reduce threats to wildlife, and avoid endangering species.

- Wildlife viewing is a leading recreational pastime on national forests and grasslands. Reductions in wildlife viewing opportunities caused by invasive species jeopardize the trust the public places in Forest Service stewards.

- Forest Service research stations have scientists who have first-hand experience in developing knowledge focused on animal species and habitats, including the adverse relationships among invasive species, animal populations, and management practices. Scientists can readily adjust their research focus to accommodate emerging issues associated with invasive issues when funding is available.

- Forest Service researchers have an extensive network of partnerships and specialized agreements and collaborations with universities, other research organizations, and multiple Federal, State, and municipal management agencies. Through existing and future collaborations, Forest Service researchers can develop knowledge that is critical for solving invasive species problems relative to terrestrial animal species.

- The Forest Service manages numerous experimental forests and ranges and multiple research natural areas that are conducive to experimental studies of wildlife and invasive species.

- The Forest Service has formed partnerships with managers of Long-Term Ecological Research Sites and National Ecological Observatory Network (NEON) Core Sites, which have the infrastructure for monitoring long-term and broad-scale trends in native and nonnative plant and animal populations. Invasive species issues are a research challenge identified by NEON.

- Many invasive species/wildlife problems are situated in urban-wildland interfaces, where urban forest institutes and ecosystem management units managed by Forest Service Research and Development can readily address them.

- Litigation over wildlife species issues produces gridlock for Forest Service managers as they attempt to implement forest and rangeland management actions. Failure to address invasive species/wildlife issues will likely lead to further gridlock.

Who Are Our Customers?

Customers seeking knowledge about the relationships between invasive species and terrestrial animal species include land managers and professionals from numerous land and natural resources agencies, including those from the National Forest System, State and Private Forestry, and International Forestry. In addition, agencies that regularly approach us to develop research studies, obtain information, or consult on best management practices include the Bureau of Land Management, USDA Agricultural Research Service and the Natural Resources Conservation Service, U.S. Fish and Wildlife Service, National Park Service, Bureau of Reclamation, Bureau of Indian Affairs, Native American Tribes, Army Corps of Engineers, Department of Defense, National Wildlife Refuges, State fish and wildlife agencies, State parks, conservation districts, city open space managers, and many more.

We supply information directly and through extension services to assist private and research ranches, managers of leased allotments, irrigators, users of water rights, city planners, and extension service customers. We offer knowledge, funds, training, jobs, and internships to students at multiple educational levels. Numerous nongovernmental organizations, such as The Nature Conservancy, National Wildlife Federation, Defenders of Wildlife, Hawks Aloft, Partners in Flight, Ducks Unlimited, and the Rocky Mountain Elk Foundation, and recreational users, such as birdwatchers, comprise a significant portion of our customer base. Professional societies, such as The Wildlife Society, Ornithological Societies of North America, and Society for Conservation Biology, use our research information for publishing purposes and to develop science-related policy. Our scientists are invited to host society meetings, give keynote talks, and contribute scientific presentations. Our publications, ideas, and models are cited extensively in the scientific literature by other researchers. Conservationists use our information when making recommendations for managing wildlife, habitats, and ecosystem services.

Invasive species are a global problem aggravated by a global market. The Forest Service needs to greatly expand its role as an international steward by reaching out to international customers and developing nations with research information about problems and solutions regarding invasive species.

Key Future Issues

Many of our customers currently are based in rural areas. We will continue to have many of the same or similar agency customers in the future, but customer demand for our research can be expected to shift from rural communities to urban populations as rural areas become urbanized and exurban populations expand into wild lands.

Advanced communication technologies and globalization of trade have rapidly expanded our range of customers, but we have not capitalized on or appropriately recognized our role as international stewards. Our Nation's increased access to world markets is paralleled by a rapid increase in natural resource problems. One of the most significant problems associated with globalization is biotic homogenization (i.e., the ecological replacement of native species with widespread exotics) (Lockwood et al. 2000). Increases in human population growth are also increasing greenhouse gas emissions, resulting in global climate change (IPCC 2007) that has the potential to alter distributions of invasive and native species.

In 15, 30, or 50 years, we may be focusing more of our time and funds on research directed toward the following:

- Protecting wildlife species from endangerment in invaded urban and urban-wildland interfaces.

- Understanding and managing plant and animal species invasions and consequent habitat losses associated with climate change and global warming.

- Managing riparian habitats and riparian-obligate animal species affected by invasions linked to water management practices such as flood control, surface water diversions, and ground water pumping.

- Managing and preserving sensitive or vulnerable ecosystems that act as uninvaded islands having high biological diversity, endemic animal species, or endangered populations of native animal species.

- Restoring wildlife habitats damaged by introduced insect species, such as Africanized honeybees (*Apis mellifera scutellata*), or degraded by invasive plant species that alter nest substrates and habitat structures.

- Restoring wildlife habitats prone to outbreaks of fire associated with plant invasions.

- Recovering endangered species affected by conversion of native habitats to monotypic vegetation composed of an alien plant species.

- Managing animal populations in fragmented, suburbanized deserts and rangelands invaded by species such as buffelgrass (*Pennisetum ciliare*) and lovegrass (*Eragrostis* spp.).

- Managing urban-wildland interface corridors used by animals that spread the seeds of invasive plants.

- Managing land fragments and habitat remnants vulnerable to plant invasions and habitat conversions that result in loss of critical wildlife habitat.

- Mitigating the effects of fragmentation-induced invasions on animal and native plant populations.

- Detecting and managing the spread of diseases and viruses carried by, infecting, or killing wildlife species.

- Detecting and eradicating or intensively managing suites of invasive and affected native species based on advanced technologic capabilities for detecting "hotspots" and native/invasive relationships (Hof et al. 2006, Stohlgren et al. 2006).

- Understanding how patterns and movement of genes and organisms across the landscape affect biodiversity, ecosystem function, and the spread of infectious diseases and invasive species.

- Understanding the role of outbreaks of insects, such as gypsy moth (*Lymantria dispar*), Africanized honey bee, and hemlock woolly adelgid (*Adelges tsugae*), in altering wildlife habitat.

The Future of Prevention and Prediction

- One of the single most important aspects of prediction and prevention is preventing new invasions in otherwise uninvaded areas, such as wilderness, national parks, and other refuges, followed by identifying other areas at high risk but currently low impact.

- Enhance our understanding of the mechanisms causing invasions that affect or involve terrestrial animal species to enable prediction of invasions and improve current risk assessments.

- Develop monitoring tools and methods for predicting and preventing the entry and emergence of harmful invasive animal species, such as Africanized honeybees, invasive rodents, brown tree snakes, java sparrows (*Padda oryzivora*), and others under the Lacey Act and listed on the Federal Register Notices.

- Develop strategies to predict and prevent the introduction and spread of animal species (e.g., rats, snakes, toads, birds) that are likely to be invasive based on knowledge of the invasive potential of introduced related species from similar climate zones.

- Likewise, develop strategies to predict and prevent the introduction and spread of invasive plants, insects, and diseases that negatively affect and endanger wildlife species and local populations based on what has been learned from earlier outbreaks of similar species.

- Enhance our understanding of attributes that make terrestrial animal communities or their natural habitats most susceptible to invasion and provide guidelines to reduce the vulnerability of communities to invasion.

- Develop treatment options to discourage nonnative species invasions following natural disturbances such as wildfires and fuel treatments.

- Model the influences of disturbance history, topography, geography, precipitation and temperature patterns, and climate change on the distributional relationships among invasions, biological diversity, and patterns of species endangerment.

- Develop models and knowledge to predict how drought, hurricanes, global warming, and fire influence outbreaks of invasive pests harmful to wildlife and their habitats.

- Communicate with the public about invasive species, such as feral pets, horticultural plants, and diseases dispersed by animals, and develop support and understanding of the importance of early prediction and prevention.

The Future of Detection and Eradication

- Develop tools and methods for detecting, prioritizing, and eradicating invasive plant and insect species that have the potential to harm or endanger wildlife species either directly by increasing mortality rates, indirectly by reducing habitat quality or availability, or broadly by reducing overall biological diversity.

- Identify priority geographical areas for treatment of invasive species based on sensitivity of wildlife species to harmful invasions or based on numbers of wildlife species that could be negatively affected.

- Develop tools and methods for detecting and eradicating harmful invasive animal species such as Africanized honeybees, invasive rodents and snakes, nutria (*Myocastor coypus*), and barred owls (*Strix varia*) that colonize new areas, where they may decimate habitats or key habitat components, or parasitize, hybridize, weaken, compete with, prey upon, kill, or replace native species.

- Develop spatial maps of occurrence of individual invasive species and concentrations of species and overlay these with maps of native animal species concentrations to determine priority locations for focusing eradication efforts.

- Develop tools, models, and protocols for detecting and monitoring new invasive species populations and their rates of spread in relation to wildlife population responses.

- Communicate and collaborate with local, State, national, and international networks to detect, monitor, manage, and mitigate invaders that have a harmful effect on threatened, endangered, and sensitive wildlife species.

The Future of Management and Mitigation

- Develop tools to prioritize invasive species for control based on the extent to which they damage habitats required by native terrestrial animal species; negatively affect native species richness and biological diversity; disrupt timing and availability of food supplies; damage trophic relationships; affect keystone species; and jeopardize sensitive, threatened, and endangered species.

- Develop, compare, or refine tools to more effectively manage invasive species populations for the purpose of restoring and improving habitats for affected native wildlife species, maintaining native biological diversity, re-establishing ecosystem linkages, and recovering threatened and endangered species.

- Develop tools to prioritize invaded areas deemed as critical to the conservation and recovery of wildlife habitats, native plant communities, and animal species at risk of local extinctions.

- Develop and evaluate tools for timing treatments to maximize efficacy and minimize side effects on nontarget native wildlife species.

- Assess and describe the relationships among plant invasions, fire, wildlife habitat use, and habitat restoration to reduce fire risk in ecosystems in which plant invasions increase fire frequency (e.g., ecosystems invaded by saltcedar (*Tamarix ramosissima*), cheatgrass (*Bromus tectorum*), buffelgrass, and Lehmann lovegrass (*Eragrostis lehmanniana*) (Cox 1999).

- Refine fundamental knowledge of population genetics and ecology of priority invasive species to reduce their effects on native animal populations.

- Communicate with the public about how invasive species are linked to the imperilment of native animal species and develop support for invasive species management and mitigation based on the public appeal for wildlife.

- Manage travel corridors and dispersal pathways to prevent the spread of animal diseases and facilitate the movement of animal vectors.

The Future of Restoration and Rehabilitation

- Develop ecologically sound restoration methods that consider genetics at population, community, and ecosystem levels of integrity and resistance to reinvasion.

- Evaluate the economic and nonmonetary costs and benefits to native species and biological diversity of restoring and rehabilitating invaded ecosystems (*sensu* Pimentel et al. 2000).

- Develop guidelines for prioritizing populations, communities, habitats, and ecosystems for restoration work to maximize efficacy and efficiency of the restoration efforts under limited resources.

- Obtain public and political support for restoration work to enhance the recovery and maintenance of wildlife habitats and animal populations, especially when the invaded ecosystem may be seen as the attractive norm.

- Develop a scientific basis for restoring and rehabilitating invaded ecosystems, considering the economic and value-added benefits to wildlife populations, recreational animal species, endangered species, and biological diversity.

- Reduce or eliminate factors in restored wildlife habitats that increase risk of reinvasion and ensure that critical components, such as food supplies and nest substrates affected by invasions, are restored for use by animal populations.

- Determine the relationships and interactions among natural disturbance regimes (e.g., fire, flooding, hurricanes, drought), species invasions, and animal survival requirements and restore disturbances found to be useful in suppressing invasions and sustaining native animals and their habitats.

The Future of Application and Communication

We need to improve our methods for disseminating research tools and information. Peer-reviewed publication of scientific results must be followed up with translation of research results into more generalized and user-friendly products. Tools should be translated into brief explanatory products that can be disseminated through brochures, Internet Web sites, and other means that more effectively reach managers and the public. Scientists need to coordinate and collaborate with State and Private Forestry and university extension services to exploit these infrastructures for more effective dissemination of important new findings and tools to a broader customer base.

We can expect increasing use and demand for our research information by multiple cultures as the ethnic composition of our Nation changes in response to growing and immigrating populations of Hispanic, Asian, and other people. Changes in human population demographics will cause shifts in our role over time. Our role will likely become increasingly oriented toward the needs and demands of urban, suburban, and exurban populations whose interests may be nontraditional, more diverse, and dictated by socioeconomic status. Differences in socioeconomic background may explain differences in how communities interact with their external environment. For example, compared with wealthier, racially mixed neighborhoods, impoverished and ethnically segregated urban areas tend to have more impoverished bird communities dominated by exotic bird species (Melles 2005). This situation is related to the amount and types of vegetation planted in neighborhoods, including whether the planted species are native and whether community planting programs are available.

We need to use advanced communication technologies to convey information about invasive species to people in developing countries. We cannot afford to isolate ourselves by ignoring our role as international stewards. Our Nation's problems with invasive species are global problems. The global spread of invasive species is leading to the worldwide impoverishment of biological diversity.

It would behoove the Forest Service to direct communication, funding, and educational efforts toward global, urban, and underserved communities, enabling them to gain an increased appreciation of and concern for the natural world as well as an understanding of the harm associated with the introduction of nonnative species. We can evolve to meet local community demands and global needs by (1) changing our workforce to match the types of customers we serve, (2) reaching out more effectively to convey information about invasive species to diverse communities and worldwide users, and (3) seeking to understand how diverse cultural, economic, immigrant, and international backgrounds can be used to guide program delivery about the economic and ecological costs of invasive species.

Top Five Priorities

1. Develop knowledge and tools to manage and mitigate effects of invasive species facilitated by globalization, urbanization, and climate change on distributions and abundances of native animal species.

Global change involves rising numbers of human immigrants and travelers, increased world trade, and advances in global communication and transportation (McNeeley 2000), all of which increase the probability that new alien pests, including plants, insects, rodents, reptiles, birds, predators, and viruses, will "hitchhike" or intentionally be released into the United States. Global change influences the scale and tempo of change in health risk pertaining to invasive viruses and diseases (McMichael and Bouma 2000). Perhaps the greatest hidden danger from invasive species is their contribution to "global homogenization," a process linked to factors ranging from communication technology to consumer mentality (McNeeley 2000). Biotic homogenization is the preferential loss of native species across taxons, or within taxonomic groups (e.g., global avifaunas), followed by ecological replacement with widespread exotics (Lockwood et al. 2000). Homogenization affects the abundance and distribution of species and the functioning of ecosystems (Collins et al. 2002). To retard the rate of global homogenization of species, future research will need to devote more attention to developing tools and methods for (1) detecting immigrating pests, viruses, and diseases that are known or have the potential to spread rapidly, jeopardizing or infecting not only humans but also native animals (including vulnerable species and disease vectors), and (2) preventing them from establishment in the United States.

Increased human activity (e.g., development) is correlated with the ecological imperilment of species (Brown and Laband 2006). Urbanization increases road density, air travel, bike and pedestrian travel, construction, and overall human activity in a given area, resulting in new pathways and wildland entry points available for dispersal of invasive species. Urbanization disturbs soil surfaces, introduces feral or escaped populations of domestic plants and animals, increases the frequency of human-caused fire outbreaks, and fragments and converts habitats. Cumulative disturbance in urban and exurban environments facilitates introductions and rapid spread of new invasive species, resulting in habitat fragmentation and reduced biological diversity (Hansen et al. 2005). Increased human population growth also accounts for increased urban warming and the development of urban "heat islands" that can attract invasive species. In addition, global rise in human activity increases the greenhouse effect, primarily through release of carbon dioxide (CO_2) emissions. Recently observed global warming is believed by many to be caused by greenhouse gas emissions from industry, transportation, and agriculture (IPCC 2007).

Patterns of species richness of native and nonindigenous plants and animals are correlated with each other and with geographical patterns of precipitation and air temperature (Stohlgren et al. 2006). Consequently, changes in climate may cause changes in the geographical distributions and concentrations of invasive and native species and may alter the susceptibility of habitats to future invasions by new nonnative species. Increasing CO_2 emissions due to human population growth can be expected to induce distributional changes in native and nonnative species either through direct effects (e.g., on photosynthetic processes) (Dukes 2000) or through effects of global warming. Global warming in combination with escalating human use of surface and ground water supplies will likely warm soils and dry waterways in some regions of the country, fostering the ability of xeric-adapted invasive species to colonize new areas and expand their ranges. The Forest Service needs to be proactive in predicting, detecting, and managing invasions and habitat losses related to changes in climate, water supply, and consumer use of water.

2. Develop knowledge and tools to improve management effectiveness and minimize unintended side effects.

The effectiveness of invasive species management and management tools (e.g., herbicides, biological control agents, fuels management) needs to be assessed for their suitability in sustaining wildlife populations, protecting biodiversity, and restoring wildlife habitats and food supplies. Most management tools cause disturbances that can facilitate invasions. All management tools, including invasive species tools, inevitably have side effects. In general, these side effects decrease with increased specificity of the management tool, but even highly specific management tools can affect native species through complex interactions (Pearson and Callaway 2003).

Invasive plant management can endanger threatened species, as in the case of the southwestern willow flycatcher, or even increase risk of human disease, as in the case of hantavirus, if not carefully applied (Dudley and Deloach 2004, Pearson and Callaway 2006). As noted by the Office of Management and Budget review of the Forest Service invasive species program, a foremost need in invasive species management now and in the future is better knowledge of our management tools and the systems we are working with to ensure that management actions improve conditions without creating more problems.

3. Develop knowledge and tools for managing invaded woodland, desert, steppe, and grassland ecosystems to conserve native species.

Old-world invasive grasses now dominate many Great Basin, Mojave, and Sonoran Deserts (Brooks and Pyke 2002), and old-world forbs are disrupting western grassland systems of the Columbia River Basin and Great Plains. Exotic grasses include the annuals cheatgrass, red brome (*Bromus rubens*), Mediterranean grass (*Schismus barbatus*), and medusahead (*Taeniatherum caput-medusae*) and the perennials buffelgrass, fountain grass (*Pennisetum setaceum*), natal grass (*Melinis repens*), and Lehmann lovegrass. These species have altered fire regimes, shortening the fire return interval. Exotic forbs include the knapweeds, leafy spurge (*Euphorbia esula*), St. Johnswort (*Hypericum perforatum*), and many others that affect wildlife populations by disrupting vegetation communities. All these invasive species thrive in post-fire landscapes.

Greater sage grouse (*Centrocercus urophasianus*) once ranged through 13 Western States and 3 Canadian provinces, but populations have declined at an overall rate of 2 percent per year from 1965 to 2003, and only about 56 percent of grouse presettlement range is currently occupied (Connelly et al. 2004). Invasive species, particularly cheatgrass and West Nile virus, pose threats to sage grouse and their habitats. Sage grouse are considered obligates of sagebrush (*Artemesia* spp.) and require large, connected landscapes of sagebrush, grasses, and forbs for their lekking, breeding, and feeding activities. Of the historical sagebrush habitat, 31 percent has been converted

to other vegetative cover, including areas invaded by alien species (Connelly et al. 2004). Cheatgrass invasion shortens the fire-return internal, reducing or eliminating fire-sensitive sagebrush (Pyke 1999). Other sagebrush bird species are also threatened by alien plant invasions (Knick et al. 2003). West Nile virus, another alien invader, represents a significant new threat to sage grouse and other at-risk bird species (DeLach 2006, Naugle et al. 2004). Research is required to develop tools and protocols for restoring and rehabilitating sage grouse habitats, reducing the amount of historic range now infested by cheatgrass, and developing measures for safeguarding sage grouse populations from infection by West Nile virus.

Spotted knapweed (*Centaurea stoebe*) and other exotic forbs have radically transformed large regions of western grasslands. Reductions in native plant abundance and diversity by these species have reduced forage for big game species and domestic livestock and eroded native food chains for songbirds by reducing invertebrate foods (Ortega et al. 2006, Trammell and Butler 1995). Buffelgrass and Lehmann lovegrass have spread throughout arid environments of Arizona. Buffelgrass chokes out native species and increases the frequency of fires in the Sonoran Desert. Fires kill native old-growth cactus, including endemic saguaro (*Carnegiea gigantea*), an important source for cavity nests of the endangered cactus ferruginous pygmy owl (*Glaucidium brasilianum cactorum*), and endemic palo verde (*Cercidium floridum, C. microphyllum*), which is used as a nest tree by many endemic bird species. Birdwatching is an important form of tourism in the Sonoran Desert environments near Tucson, where species richness of endemic and unusual bird species is remarkably high. Bock and Bock (1986) found that conversion to lovegrass communities at a Sonoran Desert site reduced numbers of species of birds, rodents, and grasshoppers. Desert tortoise (*Gopherus agassizii*) is both directly and indirectly affected by invasive plants and the fires that they cause (Brooks and Pyke 2002). In areas of recurrent fire, desert tortoise is completely absent. New studies are needed to determine the relationships among grass invasions, fire, and endemic wildlife species.

Broadleaf herbicides can be used to effectively control invasive forbs over local areas, but there is a need to better understand how best to deploy these herbicides to maximize their effectiveness at controlling target invaders and minimize their side effects on nontarget species. Some herbicides have been developed to suppress graminoids, but many are problematic due to their lack of specificity to the target invasive grass, which can result in effects on desired native grasses. Moreover, many of the serious problem grasses have achieved a scale of invasion that far exceeds effective control using herbicides. Biological control has proven effective for numerous widespread invasive forbs but less so for grasses, and biological control successes are generally sporadic. More research would advance our understanding of efficacy in biocontrol, particularly as it relates to problematic grasses. The use of grazing and fire as effective management tools for some exotic grasses and forbs should be explored further.

4. Develop knowledge and tools to manage and rehabilitate invaded riparian and wetland ecosystems, recover native species, and conserve biological diversity.

Riparian and wetland habitats have disproportionately high species richness of terrestrial animals, especially birds, relative to the percent of land area they encompass. But in many areas, these habitats are now becoming havens for many invasive species (Stohlgren et al. 1998). Invasions have the potential to impoverish the fauna inhabiting riparian and wetland communities. Even though less than 6 percent of the Earth's land mass is wetland, 24 percent of the world's most invasive plants are wetland species (Zedler and Kercher 2004). Many riparian and wetland invaders form monotypes that alter habitat structure, lower biodiversity, change nutrient cycling and productivity (often increasing it), and modify food webs. Wetlands are landscape sinks that accumulate debris, sediments, water, and nutrients, all of which facilitate plant invasions by creating canopy gaps or accelerating the growth of opportunistic plant species. Residential development and associated habitat fragmentation also increase vulnerability of riparian areas to nonnative plant invasions (Lussier and Da Silva 2005).

In the Southwest, elimination of spring flood events has reduced recruitment in cottonwood populations along many rivers and streams, allowing invading plants, such as saltcedar and Russian olive (*Elaeagnus angustifolia*), to establish on sites formerly occupied by native cottonwoods and willows. As surface water availability declines, native riparian plants senesce and invasive plants replace them. Under these conditions, desert bighorn sheep (*Ovis canadensis nelsoni*) populations and other native wildlife generally decline (Lovich and de Gouvenain 1998).

Accumulation of woody debris, combined with dense stands of invasive woody plants in the understory, has led to fuel

loadings capable of supporting catastrophic wildfires (Busch 1995). Sensitive and endangered species such as southwestern willow flycatcher (*Empidonax traillii extimus*) and yellow-billed cuckoo (*Coccyzus americanus*) depend on the presence of riparian vegetation. Fires destroy their nests, lowering their local productivity and recruitment rate. Fires destroy cottonwoods used by riparian cavity-using species, such as *Myotis* bats, woodpeckers, kestrels, Bewick's wren (*Thryomanes bewickii*), ash-throated flycatcher (*Myiarchus cinerascens*), and nuthatches, and platform nesters, such as owls, buteos, and accipiters, which require large tree species to build nests and reproduce. Invasive woody species are unsuitable as nest sites for these animals. Fire kills cottonwoods, resulting in early emergence of a critical food source (cicadas) for birds and other wildlife (Smith et al. 2006). Fire facilitates replacement by invasive plants. Managers lack information on the interactive and long-term effects of invasive plants, fires, and flood control on sensitive wildlife species and their habitats.

In California and other subtropical regions, giant reed (*Arundo donax*), a nonindigenous perennial grass, aggressively invades riparian areas, changing vegetation structure, reducing availability of perch and nest sites, and reducing numbers, total biomass, and taxonomic richness of aerial insect species (Herrera and Dudley 2003). Alteration of food supply reduces the habitat value of riparian areas to bird species and other animals whose diets are largely composed of insects found in native riparian vegetation. Endangered species obligated to riparian zones include least Bell's vireo (*Vireo bellii pusillus*) and southwestern willow flycatcher. Both species are threatened by habitat loss caused by giant reed invasion.

Land managers need increased understanding and improved tools to deal effectively with the complexity of interacting problems created by invasions in riparian ecosystems. Current restoration and rehabilitation methods for riparian areas are often not compatible with goals for recovering endangered animal species or conserving species diversity. New research could provide alternative approaches for managing riparian ecosystems to enable conservation of animal species.

5. Develop knowledge and tools to manage invasive and native species and habitats in islands and island ecosystems.

Because island faunas have evolved in isolation, they are especially sensitive to invasive exotic species of competitors, predators, and parasites to which the island dwellers have few or no defenses. Introductions of rats, dogs, cats, pigs, goats, and mongoose (*Herpestes javanicus*), as well as other animals associated with human colonization, have caused extinctions and still continue to threaten Pacific and Caribbean island species. On the island of Guam, 9 of 11 forest-dwelling bird species were extirpated following the arrival of the brown tree snake in the early 1960s. Similarly, Hawaii's endemic bird populations were extirpated in the lowlands in the 1880s as the result of the accidental introduction of mosquitoes that transmitted bird pox and avian malaria to which the natives lacked immunity.

Invasive exotic plants can also negatively affect island wildlife populations, especially those that change the environment they invade. A typical example is that of the invasive exotic grasses, which are highly susceptible to wildfires and change the fuel load such that intensive wildfires become more likely. The grasses recover quickly from fires in contrast to woody vegetation that recovers slowly, if at all, and the resulting wildfire cycle impedes forest regeneration. This type of problem is of special concern in Hawaii and Puerto Rico, where forest areas are limited and vulnerable to additional anthropogenic and natural (e.g., hurricanes) disturbances, further stressing threatened and endangered island wildlife (e.g., Puerto Rican nightjar (*Caprimulgus noctitherus*), yellow-shouldered blackbird (*Agelaius xanthomus*). Other invasive exotic plants on islands, such as Australian pine (*Casuarina equisetifolia*), Brazilian pepper (*Schinus terebinthifolius*), and *Leucaena,* are quick to colonize newly disturbed sites and can displace and dominate native early successional plant species of importance to wildlife. Such invasions are of concern for the endangered neotropical migrant Kirtland's warbler (*Dendroica kirtlandii*) on its island wintering grounds, where invasive exotic species displace native plants bearing fruit consumed by the warbler.

Island ecosystems are often easily invaded and colonized by exotic species, in part because island biotas are poorly adapted to compete with or evade newly arriving species. A recent example comes from Hawaii, where the coqui frog (*Eleutherodactylus* spp.) introduced from Puerto Rico has successfully established lowland populations that now have densities three times higher than those found in Puerto Rico, presumably due to the absence of coqui predators on Hawaii. The success of exotic colonists on islands is evident in Puerto Rico's resident breeding bird species of which 25 percent are alien or exotic species: 31 species of exotic birds are established as breeders; 5

exotic species are found in the wild, but breeding has not been established. Exotic birds pose a threat to native species because they have the potential to transmit diseases directly; serve as a reservoir for diseases transmitted by arthropod vectors; and/or elevate pathogen levels, enabling them to persist in higher than normal concentrations.

The threat of disease transmission from an exotic bird species is of special concern for the endangered Puerto Rican parrot (*Amazona vittata*), which is a species for which high disease susceptibility is predicted. The recent (2005) evidence for the mosquito-transmitted West Nile virus in birds on Puerto Rico demonstrates that research to more effectively predict, detect, and manage such threats to the parrot and other endangered species is of critical importance.

Another newly arrived invasive exotic to threaten the Puerto Rican parrot is the Africanized honeybee that appeared in the 1990s. The Africanized honeybees have hybridized with the previously naturalized European honeybees (*Apis mellifera*), resulting in a more aggressive colonizer of tree cavities used for nesting by parrots and other wildlife. The potential exists to deter bee colonization of nest cavities, as preliminary research by the International Institute of Tropical Forestry indicates that pheromones may prove useful as a deterrent to bee colonization.

Although long established as exotics since the arrival of Europeans, rats continue to threaten island wildlife throughout the world, and control programs continue to this day. Rats are also predators of threatened and endangered frogs, lizards, and snakes, including the Puerto Rican (*Epicrates inornatus*) and Virgin Island (*E. monensis granti*) boas. In addition to the direct effect that black rats may have on island wildlife, they may indirectly affect wildlife by changing forest plant composition, as a result of their consumption of seeds. Despite its potential importance, the role of black rats (*Rattus rattus*) as seed predators influencing tree recruitment and subsequent forest composition has yet to be studied in island or tropical ecosystems.

In summary, because of the high rates of colonization and establishment of exotic species facilitated by human activities, island ecosystems are ideal laboratories for studying invasive exotics and their potential effects. These studies are required for the recovery of endangered island species, and, in many instances, the findings from such studies are of relevance to the study and management of invasive exotics elsewhere, such as in the Southern United States.

For similar reasons, other types of isolated "island-like" ecosystems are also at greater risk to effects of invaders than "mainland" systems. Examples include sky islands resulting from mountains in desert environments, islands in large lakes and river systems, and habitat patches isolated by development. In each case, unique (in some cases, endemic) fauna may experience greater threats from invasions, in part, because they may be associated with small populations and small habitat areas that are surrounded by potential invaders.

Literature Cited

Bock, C.E.; Bock, J.H. 1986. Ecological effects of planting African lovegrasses in Arizona. National Geographic Research. 2: 456–463.

Brooks, M.L.; Pyke, D.A. 2002. Invasive plants and fire in the deserts of North America. In: Galley, K.E.M.; Wilson, T.P., eds. The role of fire in the control and spread of invasive species. Misc. Publ. No. 11. Tallahassee, FL: Tall Timbers Research Station: 1–14.

Brown, R.M.; Laband, D.N. 2006. Species imperilment and spatial patterns of development in the United States. Conservation Biology. 20: 239–244.

Busch, D.E. 1995. Effects of fire on southwestern riparian plant community structure. Southwestern Naturalist. 40: 259–267.

Collins, M.D.; Vazquez, D.P; Sanders, N.J. 2002. Species-area curves, homogenization and the loss of global diversity. Evolutionary Ecology Research. 4: 457–464.

Cox, G.W. 1999. Alien species in North America and Hawaii: impacts on natural ecosystems. Washington, DC: Island Press. 387 p.

DeLach, A. 2006. Invasive species in the northwestern United States: threats to wildlife, and defenders of wildlife's recommendation for prevention policies. Northwestern Naturalist. 87: 43–55.

Dudley, T.L.; DeLoach, C.J. 2004. Saltcedar (*Tamarix* spp.), endangered species, and biological weed control—can they mix? Weed Technology. 18: 1542–1551.

Dukes, J.S. 2000. Will the increasing atmospheric CO_2 concentration affect the success of invasive species? In: Mooney, H.A.; Hobbs, R.J., eds. Invasive species in a changing world. Washington DC: Island Press: 95–114.

Hansen, A.J.; Knight, R.L.; Marzluff, J.M., et al. 2005. Effects of exurban development on biodiversity: patterns, mechanisms, and research needs. Ecological Applications. 15: 1893–1905.

Herrera, A.M.; Dudley, T.L. 2003. Reduction of riparian arthropod abundance and diversity as a consequence of giant reed (*Arundo donax*) invasion. Biological Invasions. 5: 167–177.

Hof, J.; Flather, C.; Baltic, T.; King, R. 2006. Nonparametric projections of forest and rangeland condition indicators: a technical document supporting the 2005 USDA Forest Service RPA Assessment Update. Gen. Tech. Rep. RMRS-GTR-166. Fort Collins, CO: U.S. Department of Agriculture, Forest Service, Rocky Mountain Research Station. 39 p.

Intergovernmental Panel on Climate Change (IPCC). 2007. Climate change 2007: the physical science basis. Contribution of Working Group I (Solomon, S.; Qin, D.; Manning, M., et al., eds.) to the fourth assessment report of the Intergovernmental Panel on Climate Change. New York: Cambridge University Press., http://www.ipcc.ch/ipccreports/ar4-wg1.htm. Accessed January 10, 2010.

Knick, S.T.; Dobkin, D.S.; Rotenberry, J.T., et al. 2003. Teetering on the edge or too late? Conservation and research issues for avifauna of sagebrush habitats. The Condor. 105: 611–634.

Lockwood, J.L.; Brooks, T.M.; McKinney, M.L. 2000. Taxonomic homogenization of the global avifauna. Animal Conservation. 3: 27–35.

Lovich, J.E.; de Gouvenain, R.G. 1998. Saltcedar invasion in desert wetlands of the Southwestern United States: ecological and political implications. In: Majurdar, S.K.; Miller, E.W.; Brenner, F.J., eds. Ecology of wetlands and associated systems. Easton, PA: Pennsylvania Academy of Science: 447–467.

Lussier, S.M.; Da Silva, S.N. 2005. Plant invasions in Rhode Island riparian areas. Rhode Island Naturalist. 12: 1–24.

Mack, R.N.; Simberloff, D.; Lonsdale, W.M., et al. 2000. Biotic invasions: causes, epidemiology, global consequences and control. Issues in Ecology. 5: 1–20.

McMichael, A.J.; Bouma, M.J. 2000. Global changes, invasive species, and human health. In: Mooney, H.A.; Hobbs, R.J., eds. Invasive species in a changing world. Washington, DC: Island Press: 191–210.

McNeeley, J.A. 2000. The future of alien invasive species: changing social views. In: Mooney, H.A.; Hobbs, R.J., eds. Invasive species in a changing world. Washington, DC: Island Press: 171–190.

Melles, S.J. 2005. Urban bird diversity as an indicator of human social diversity and economic inequality in Vancouver, British Columbia. Urban Habitats. 3: 25–48.

Naugle, D.E.; Aldridge, C.L.; Walker, B.L., et al. 2004. West Nile virus: pending crises for greater sage-grouse. Ecology Letters. 7: 704–713.

Ortega, Y.K; McKelvey, K.S.; Six, D.L. 2006. Invasion of an exotic forb impacts reproductive success and site fidelity of a migratory songbird. Oecologia. 149: 340–351.

Pearson, D.E.; Callaway, R.M. 2003. Indirect effects of host-specific biological control agents. Trends in Ecology and Evolution. 18(9): 456–461.

Pearson, D.E.; Callaway, R.M. 2006. Biological control agents elevate hantavirus by subsidizing mice. Ecology Letters. 9: 442–449.

Pimental, D.; Lach, L.; Zuniga, R.; Morrison, D. 2000. Environmental and economic costs of nonindigenous species in the United States. Bioscience. 50: 52–65.

Smith, D.M.; Kelly, J.F.; Finch, D.M. 2006. Cicada emergence in southwestern riparian forest: influences of wildfire and vegetation composition. Ecological Application. 16: 1608–1618.

Stohlgren, T.J.; Bardeau, D.; Flather, C., et al. 2006. Species richness and patterns of invasion in plants, birds, and fishes in the United States. Biological Invasions. 8: 427–447.

Stohlgren, T.J.; Bull, K.A.; Otsuki, Y., et al. 1998. Riparian zones as havens for exotic plant species. Plant Ecology. 138: 113–125.

Trammell, M.A.; Butler, J.L. 1995. Effects of exotic plants on native ungulate use of habitat. Journal of Wildlife Management. 59: 808–816.

Wilcove, D.S.; Rothstein, D.; Dubow, J., et al. 2000. Leading threats to U.S. biodiversity: What's threatening imperiled species? In: Stein, B.A.; Kutner, L.S.; Adams, J.S., eds. Precious heritage: the status of biodiversity in the United States. New York: Oxford University Press USA: 238–254.

Wilson, E.O. 2002. The future of life. New York: Alfred A. Knopf. 256 p.

Zedler, J.; Kercher, S. 2004. Cause and consequences of invasive plants in wetlands: opportunities, opportunists, and outcomes. Critical Reviews in Plant Science. 23: 431–452.

The Role of the Forest Service in Aquatic Invasive Species Research

Susan B. Adams[1], Kelly M. Burnett[2], Peter Bisson[3], Bret Harvey[4], Keith H. Nislow[5], Bruce E. Rieman[6], and John Rinne[7]

Abstract

Aquatic ecosystems include the most imperiled taxa in the United States, and invasive species are the second leading contributor to this imperilment. The U.S. Department of Agriculture (USDA), Forest Service is legally mandated to sustainably manage aquatic habitats and native species on National Forest System (NFS) lands. Invasive species add complexity and uncertainty to natural resource management, and, thus, invasive species research is needed to guide effective, science-based management of aquatic systems. Although Forest Service Research and Development (R&D) scientists have much expertise to apply, aquatic invasive species research has not been an agency focus. We identify areas in which the Forest Service is well positioned to contribute research that other organizations are not addressing. Increasing agency emphasis on aquatic and riparian invasive species research and adding expertise in several areas (e.g., risk assessment, genetics, and several taxonomic areas) would facilitate a shift toward the Forest Service providing more valuable science and leadership in this arena. We identify some key general research needs; however, a more formal process, bringing Forest Service aquatic and riparian scientists together, perhaps with key NFS biologists and other stakeholders, is necessary to effectively identify and prioritize specific research needs. Some of the top research needs we identify include the following:

- Develop new prediction and ecological risk assessment tools and conduct risk assessments for priority invasive species and habitats.

- Collaborate on or establish a central data management repository.

- Increase understanding of ecological, physical, and biological factors facilitating and inhibiting invasions.

- Develop new prevention, eradication, and control tools.

- Enhance role of social sciences in aquatic invasive species research.

- Improve communications. Bring Forest Service R&D scientific expertise to bear on aquatic invasive species policy and regulation. Improve communication with NFS and other biologists and the public.

Importance of Aquatic and Riparian Invasive Species

Aquatic and riparian-associated species constitute the Nation's most imperiled biota, with the five most imperiled groups residing in freshwater and riparian habitats (fig. 1). Invasive species are the second most important factor in this imperilment, contributing to the declines of about one-half of the imperiled species (fig. 2). Invasive species can harm native communities via competition, predation, hybridization, and habitat alteration and as sources and vectors of alien pathogens. Species invasion is a global problem, and an international perspective is necessary to effectively address many invasion issues.

[1] Research Fisheries Scientist, Forest Service, Southern Research Station, Center for Bottomland Hardwoods Research, 1000 Front Street, Oxford, MS 38655.

[2] Research Fisheries Biologist, Forest Service, Pacific Northwest Research Station, Aquatic and Land Interactions Program, 3200 SW Jefferson Way, Corvallis, OR 97331.

[3] Research Fisheries Biologist, Forest Service, Pacific Northwest Research Station, Aquatic and Land Interactions Program, 3625 SW 93rd Avenue, Olympia, WA 98512.

[4] Research Fisheries Biologist, Forest Service, Pacific Southwest Research Station, Watershed Project, 800 Buchanan Street, Arcata, CA 95521.

[5] Research Fisheries Biologist, Forest Service, Northern Research Station, Fish and Wildlife Habitat Relationships in New England Forested Ecosystems, 201 Holdsworth NRC, University of Massachusetts, Amherst, MA 01003.

[6] Emeritus Research Fisheries Scientist, Forest Service, Rocky Mountain Research Station, Boise Aquatic Sciences Laboratory, 322 East Front Street, Suite 401, Boise, ID 83702.

[7] Research Fisheries Biologist, Retired, Forest Service, Rocky Mountain Research Station, SW Forest Science Center, 2500 South Pine Knolls Drive, Flagstaff, AZ 86001.

Figure 1.—*Degree of imperilment of various plant and animal groups (redrawn from Master et al. 2000).*

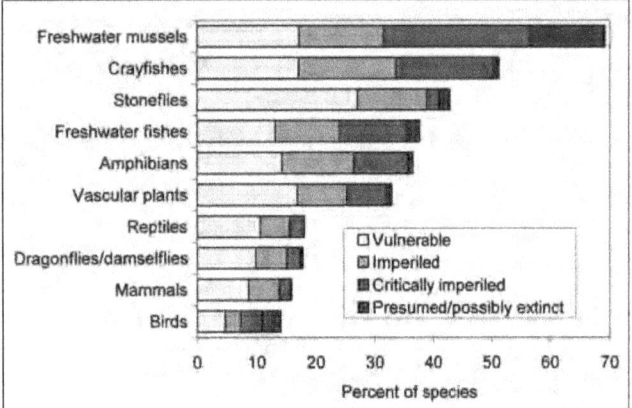

Figure 2.—*Causes of imperilment of imperiled and federally listed species in the United States (redrawn from Wilcove et al. 2000).*

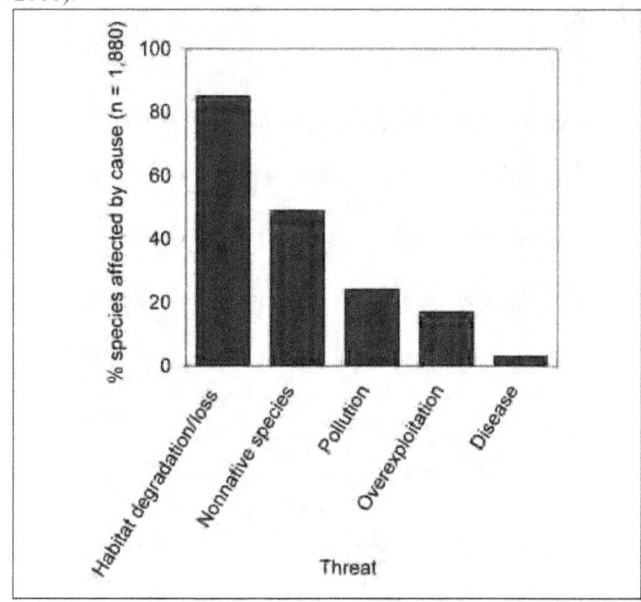

Forest Service Mandate To Address Aquatic and Riparian Invasive Species Issues

The external panel (hereafter, "the Panel") charged with reviewing the Forest Service Research and Development (R&D) Invasive Species Strategic Program Area stated, "The mandate of the Forest Service (FS) and its current commitment to management of aquatic habitats is unclear from the briefing materials" (Raffia et al. 2006). Although perhaps not articulated to the Panel, a clear legal mandate for the Forest Service to manage aquatic habitats is conferred by three key laws—the Forest Service Organic Administration Act of 1897, the Multiple-Use Sustained-Yield Act of 1960, and the National Forest Management Act of 1976. These laws state that national forests are to be established and administered to secure favorable conditions for water flows and to provide the American people with multiple uses and sustained yields of renewable resources, including those related to watersheds, wildlife, and fish. Furthermore, under the Clean Water Act of 1977 and the Endangered Species Act (ESA) of 1973, Federal agencies are to "restore and maintain the chemical, physical, and biological integrity of the Nation's waters" and to ensure that actions they "authorize, fund, or carry out must not jeopardize the continued

existence of any listed species or result in the destruction or adverse modification of critical habitat." USDA policy directs the Forest Service to "maintain viable populations of all native wildlife, fish, and plant species in habitats distributed throughout their geographic range on NFS lands" (USDA Forest Service 1995).

Forest Service R&D is directed to provide technical assistance to the National Forest System (NFS) in meeting its legal mandates as well as to other managers (other Federal agencies, tribes, States, and private landowners) of the Nation's 731 million acres of forested lands. As manager of 192 million acres of national forests and grasslands, which include 2 million acres of lakes, 300,000 miles of perennial streams, 200,000 miles of fishable streams, and 42 million acres of municipal watersheds, the Forest Service can influence the introduction, establishment, and spread of aquatic and riparian invaders through its policies, as well as by leadership in habitat management actions to control unwanted invaders. In the Western United States, roughly 75 percent of all water originates on NFS lands; thus, the Forest Service has a strong influence on the Nation's water resources. Water issues are a prominent and increasing part of the agency's interests, and invasive species can have a major influence on water quality, water availability, and aquatic biological integrity. Following are some regional examples of the threats that aquatic and riparian invasive species pose to the Nation's aquatic resources on Federal lands.

The *Interior Columbia River Basin* has 88 recognized native taxa of fishes and 55 nonnative taxa (Lee et al. 1997). Two introduced fishes are now the most widely distributed of any fish taxa (native or nonnative) in the basin. Roughly one-half of the native fishes are of conservation concern, due in large part to invasive species. Large, warm, low-elevation habitats are among the most invaded aquatic communities, but invasions continue to progress upstream. In many cases, spread of invasive species is facilitated by human activities (e.g., habitat alteration and fish stocking), but climate change and shifting hydrologic processes may extend or accelerate the process. In addition, many high-elevation lakes have been stocked with nonnative fishes.

In the *Southwestern United States*, reservoir construction and fisheries management have contributed to an irruption of aquatic invasions and subsequent imperilment of native fishes (Rinne 1996). For example, during the 20th century, more than one-half of the 100 nonnative fish species introduced in Arizona became established. Due to extensive modification of low-elevation rivers, most remaining perennial, riverine habitats are on NFS lands and serve as refugia for native species, 70 percent of which are listed under the ESA (Rinne and Medina 1996).

In the *South Atlantic-Gulf of Mexico* region (east of the Mississippi River) the U.S. Geological Survey (USGS) Nonindigenous Aquatic Species database indicates established populations of 226 nonnative aquatic species, 122 of which are not native to North America (USGS 2008). Nonnative sport fishes have displaced related, native species. Some of the 52 nonnative aquatic plant species have created major habitat changes and altered biological and physical ecosystem functions, leading to native species displacement and loss of recreational and commercial opportunities among other effects.

The *Great Lakes Basin* has more than 180 established, nonnative aquatic species and has the highest known invasion rate for a freshwater ecosystem; a new invader is discovered every 28 weeks, on average (Ricciardi 2006). High-profile species (e.g., sea lampreys and zebra mussels) cause large economic losses, but many other, less-publicized species also cause substantial ecological disruptions.

In the *Northeastern United States*, aquatic and riparian invasive species pose problems for major river and species restoration efforts. Nonnative fishes, in combination with habitat change, complicate the conservation and restoration of a suite of native diadromous fishes, including the last wild Atlantic salmon populations (listed under ESA) in the United States (National Research Council 2004). Invasive riparian plants (e.g., purple loosestrife and Japanese knotweed) threaten the success of ecological flow prescriptions designed to restore threatened floodplain forest communities (Nislow et al. 2002).

Roles of Forest Service Research and Development in Aquatic and Riparian Invasive Species Research

Past and Current Roles

The Forest Service R&D has not been a national leader in aquatic invasive species research, although Forest Service R&D scientists have conducted excellent research on some aquatic invasions. Despite the enormous threats invasive species pose to aquatic ecosystems, invasive species research within Forest Service R&D has focused primarily on weeds, insects, and diseases harmful to forests and rangelands. Overall, momentum and funding for Forest Service research on invasive aquatic and riparian species lag far behind those for invasive upland species.

Given the agency's lack of emphasis on and funding for invasive aquatic and riparian research, potential near-term program strengths are not necessarily reflected in research to date. The Forest Service R&D maintains a strong group of fish and aquatic ecologists who are well qualified to conduct invasive species research; indeed, most have researched invasive species at some time in their careers. Thus, if agency funding priorities were directed toward such work, Forest Service scientists would be well positioned to conduct research addressing aquatic invasive species.

Future Roles

The Panel concluded that Forest Service R&D on invasive species needs to be strengthened, administrative burdens on scientists reduced, and an independent scientific board established to advise administrators at the national level. We concur. However, we strongly disagree with the Panel's reluctance

"...to recommend increases in investment in aquatic habitat management at the risk of further weakening the traditional terrestrial-related FS research" (Raffa et al. 2006: 10).

Although other Federal agencies share responsibilities for addressing aquatic and riparian invasions, their management and research priorities generally differ from those of the Forest Service (table 1), making it essential for Forest Service R&D to participate in future research in this area. In one ongoing effort to prioritize risks from invasive species, for example, the lowest level of threat to which invaders are assigned is a "threat to ecosystem health" (NISC and ANSTF 2007); however, to land managers, ecosystem health is a high priority. Because of different priorities, minimal research can be expected by other Federal agencies on aquatic and riparian invasive species in headwater and high-elevation rivers where many public lands occur, an area in which Forest Service R&D expertise is recognized internationally. Further, in headwater systems, where terrestrial and aquatic ecosystem contact is maximized, the probability of establishment and effects of invasive species are likely to be most dependent on land management and upland habitat changes.

The Forest Service is in a position to conduct invasive species research over longer periods than the graduate student cycle typified by most academic research programs. Long-term research is necessary for developing control and eradication tools, monitoring effectiveness of control efforts, learning by adaptive management, and understanding how long-term changes (e.g., in climate or fire regimes) influence the spread and effects of invasive species. Long-term research is particularly relevant in aquatic systems because aquatic population and habitat responses to many land management actions may lag significantly behind those in terrestrial systems. Thus, identifying influences of upland habitat alteration on the vulnerability of aquatic habitats to invasion requires long-term study.

Forest Service R&D is poised to address aquatic and riparian invasive issues on regional, national, and international scales through long-term partnerships that scientists have established

Table 1.—*Roles of other selected Federal agencies in aquatic invasive species research and management.*

Program	Area of invasive species responsibility
U.S. Department of the Interior	
U.S. Fish and Wildlife Service, ANSTF[1]	Cochair of ANSTF. Employs an ANS Coordinator in each region. Directs funding to regional ANSTF panels. Works with States to develop State ANS plans.
USGS, Science Centers and Cooperative Education Units	Conducts research on invasive aquatics and riparian plants, but focuses primarily on large rivers and wetlands at low elevations. Includes research support for Department of Interior lands. Maintains national invasive species databases.
Bureau of Reclamation	Conducts research on control and monitoring of aquatic and riparian invasive species, primarily in large, regulated rivers.
NOAA	
ISP	Cochair of ANSTF. Supports research on invasive species issues related to marine, estuarine, and diadromous organisms and introductions via marine pathways, such as ballast water introductions.
NCRAIS	Fosters cross-NOAA leadership, communication, and coordination for NOAA's research investments in support of understanding, preventing, responding to, and managing aquatic species invasions in U.S. coastal ecosystems (including the Great Lakes ecosystem).
USDA APHIS	
National Wildlife Research Center	Conducts research on aquatic invasive species but places priorities on birds, mammals, wildlife diseases, and aquatic plants.
U.S. Department of Defense	
U.S. Army Corps of Engineers	Conducts research and control studies on aquatic invasive plants.

ANS = aquatic nuisance species. ANSTF = Aquatic Nuisance Species Task Force. APHIS = Animal and Plant Health Inspection Service. ISP = Invasive Species Program. NCRAIS = National Center for Research on Aquatic Invasive Species. NOAA = National Oceanic and Atmospheric Administration. USDA = U.S. Department of Agriculture. USGS = U.S. Geological Survey.

[1] "The Aquatic Nuisance Species (ANS) Task Force is an intergovernmental organization dedicated to preventing and controlling aquatic nuisance species, and implementing ... the National Invasive Species Act (NISA) in 1996. The task force consists of 10 Federal agency representatives and 12 Ex-officio members. The task force coordinates governmental efforts dealing with ANS in the United States with those of the private sector and other North American interests via regional panels and issue specific committees and work groups" (USFANSTF n.d.). Although not a member of the task force, the Forest Service is a member of some regional panels and committees.

with (1) NFS biologists and botanists; (2) the national network of Forest Service research natural areas and experimental forests and ranges; (3) Forest Service International Programs; and (4) scientists working at long-term ecological research sites, National Ecological Observatory Network (NEON) sites, and other agencies, organizations, and universities. Because of its broad geographic scope, Forest Service R&D is in an excellent position to study species that are native in some parts of the United States and invaders in others (e.g., brook trout, rainbow trout, Atlantic salmon, and red swamp crayfish).

Customers for Forest Service Research on Aquatic and Riparian Invasive Species

The Forest Service aquatic research program, including invasive species research, has a broad customer base. Any organization or person interested in the conservation of aquatic biological diversity, aquatic ecological function, recreational and commercial fishing, and interactions between land management and invasive species is a potential customer.

Government customers in the United States (including Federal, State, local, and tribal natural resource management and regulatory agencies and other government entities and politicians) use Forest Service R&D aquatic invasive species research findings to inform decisions related to fish stocking, species conservation, and habitat management. Aquatic invasive species research by the Forest Service is also conducted and used internationally where similar invasion issues exist (e.g., rainbow trout in South America); international collaborations can be critical to our understanding of invasions in the United States (e.g., expertise from Australia and Central America regarding waterborne chytrid fungus causing mass mortality of amphibians in the United States).

All Forest Service branches (International Programs, State and Private Forestry [SPF], and NFS) use Forest Service R&D results on aquatic invasive species. International Programs, SPF, and Forest Service technology transfer professionals use the research results to inform landowners, recreationists, and other parties how to recognize aquatic and riparian invasive species and how to help minimize spread. Given that introduced species can arrive and spread on NFS lands via

recreational activities, a better-informed public is essential to reducing spread of invasive species. NFS biologists and botanists have clear and immediate needs for Forest Service research on invasive species, both to manage ongoing invasions and to prevent future invasions. The latter is important because the introduction and spread of some aquatic and riparian invasive species have been facilitated by NFS activities such as road building, timber harvest in riparian areas, reforestation, firefighting (water transfers), erosion control measures (e.g., seeding or planting nonnative plants), stream and riparian restoration, stocking nonnative game fish, and providing motorized recreation access (e.g., campgrounds and boat launches).

Nongovernmental users of Forest Service R&D research include academic scientists, nongovernmental organizations focused on aquatic conservation or natural resource management, professional societies, fishing clubs, and the public, including rural communities. A shift to more urbanized customers may alter demands on natural resources, expanding emphases on clean water and nonconsumptive recreational uses from NFS lands and influencing research priorities on invasive species.

Key Future Aquatic and Riparian Invasive Species Research Issues

The key future research issues regarding aquatic and riparian invasive species fall into two general categories: (1) questions directly related to prevention, prediction, management, etc., of invasive species (addressed under subheadings below), and (2) conservation and ecosystem management questions to which threats from invasive species add new complexity. The latter includes understanding how invasions influence the probability of persistence for native species, defining the potential roles of public lands in providing refugia for native aquatic communities, and identifying and addressing conflicting resource management goals related to aquatic invasions.

Federally managed lands serve as refugia for many species. Aquatic and riparian invasions have typically proceeded upstream from low-elevation habitats and so have contributed substantially to the functional fragmentation of aquatic networks. Consequently, many native species persist only in isolated remnant populations in headwater systems, intensifying their susceptibility to extinction. As fragmentation increases, forest management will become even more critical to the conservation of remnant aquatic biological diversity. Identifying aquatic refugia and prioritizing them for protection is part of many conservation-planning processes. Identifying refugia that are less prone to invasion is important, because these areas will become the strongholds of biotic integrity and sources of colonists to repopulate newly restored, connected habitats. More research on invasion mechanisms of particular species and on habitat and biotic characteristics conferring resistance to invasion would facilitate conservation planning. We predict an increased need for Forest Service research related to managing habitats and populations fragmented by nonnative species. Many requests for invasive species research are driven by the ESA (e.g., threats of invaders to ESA-listed aquatic fauna), which will continue to strongly influence management and, thus, research priorities.

Many existing aquatic invasive species issues stem from conflicting goals and values within and among agencies and the public. Conflicts arising before introduction of a species (table 2) often relate to differences in values, assessments of risk, or willingness to accept invasion risk. For example, goals of game fish stocking to provide recreational fishing opportunities in wilderness lakes may conflict with goals of conserving

rare species or of maintaining areas where natural processes predominate. Conflicts over management goals also arise after invasion (table 3) but may not be immediately apparent. The potentially complex tradeoffs between preempting and allowing nonnative trout invasions in mountain rivers illustrate this point. Constructing barriers to preempt invasion by a nonnative trout can isolate native trout populations, eliminating the expression of migratory life histories that may be key to their long-term persistence. Thus, society must sometimes choose between an isolated population of a native species that depends on active management for persistence or a nonnative form that may retain more resilience and fill a similar ecological role. Although scientific understanding alone will not resolve conflicts, it can help society answer the tough questions related to invasive species issues. Forest Service R&D has and can continue to provide science that informs the discussion of conflicting goals by distinguishing facts from values in the decisionmaking process, illuminating conflicting goals, and predicting outcomes, risks and tradeoffs of various management activities related to aquatic invasions. Forest Service R&D research on high mountain lake fish stocking provides a good model of research constructively contributing to addressing conflicting goals.

Table 2.—*Examples of conflicting goals with respect to human activities potentially leading to intentional species introductions.*

Goals favoring native ecosystems	Goals favoring potentially invasive species
Conserve native biodiversity/ecological integrity (conservation of threatened and endangered species).	Recreation—game fish stocking, use of live bait, motorized vehicle/boat access.
Wilderness values—maintenance of natural processes (includes legal mandates).	Agriculture—species importation for aquaculture or for control of other organisms (e.g., mollusks in fish farming), live food trade.
Protection of existing commercial interests (e.g., commercial fishing for native species or tourism).	Commerce—importation or transfer of species for pet and nursery trades.

Table 3.—*Examples of conflicting resource management goals after invasion.*

Eradicate or control invasive species		Maintain or promote invasive species
Restore native community.	←→	Ongoing fish stocking for recreation.
Install barriers to halt invasion—persistence of native species only with active management.	←→	Maintain/restore connectivity and allow invasion—persistence of nonnative with similar ecological function without need for active management.
Use of chemicals to eradicate invasive species—risk to some nontarget native organisms, public health concern.	←→	No chemical use—persistence of nonnative, reduced risk to some nontarget organisms.
Economic—restore commercial interests based on native species (commercial harvest, tourism, etc.).	←→	Economic—maintain economic value from invasive species (aquaculture, recreation).

Prevention and Prediction

Although prevention is not easily tallied in annual accomplishment reports, it is the most effective tool for countering invasive species. Prevention of aquatic invasive species requires attention at three scales: (1) keeping new invasive species out of North America, (2) preventing invasions across natural boundaries (e.g., among river basins), and (3) preventing the spread of invasive species to new habitats within river basins. Forest Service R&D has played a role in the latter by researching and providing guidance to managers on the tradeoffs of installing instream barriers to prevent upstream invasions. Also, Forest Service R&D participated in developing effective methods and guidelines for cleaning equipment used in firefighting and other activities to limit spread of aquatic invasive species. Forest Service R&D can play a bigger role in the first two scales both via research (e.g., risk assessments) and by informing policy decisions made by other agencies (e.g., the Animal and Plant Health Inspection Service and the U.S. Fish and Wildlife Service) on intentional importation and interdrainage transfers of live aquatic and riparian organisms. The Forest Service participates in some networks addressing such issues (e.g., GLRC 2005: appendix A), but coordinated efforts vary regionally.

Prediction of possible outcomes resulting from nonnative aquatic species invasions is a high priority need that Forest Service R&D can help meet by developing ecological risk assessment tools. Risk assessment components are most useful when developed in the context of an interactive, computer-based decision support system that can be readily accessed by risk assessors and risk managers to (1) describe and understand the current distributions (sources) of aquatic invasive species; (2) predict the future establishment, spread, and consequences of aquatic invasive species based on species characteristics, aquatic habitat and biotic community characteristics, and potential pathways of spread; (3) identify locations where control technologies may efficiently limit the spread of aquatic invasive species; and (4) evaluate the overall effectiveness and net benefits afforded by alternative control measures proposed for specific locations. Risk assessments on aquatic invasive species for NFS lands are likely to include local vectors more than the international trade and transportation vectors emphasized as key sources of introduced aquatic organisms to the continental United States (Lodge et al. 2006). Improved understanding of invasions in the ecological context of habitat conditions (e.g., Brown and Moyle 1997, Harvey et al. 2004)

will be critical for effective risk assessments. A long-term strength of Forest Service R&D has been discovering the ecological roles of natural and anthropogenically influenced disturbance regimes (fires, floods, insect outbreaks, etc.). Therefore, Forest Service R&D is well positioned to address how modification of disturbance regimes influences invasion probability and susceptibility.

Detection and Eradication

The Panel accurately noted that biological monitoring on NFS lands is weak for early identification of aquatic and riparian invasions. Few, if any, national forests have monitoring programs designed to detect an array of potential aquatic or riparian invaders. The probability of detecting aquatic nonnatives early in an invasion varies by region, depending on the number of aquatic and riparian biologists working in the area, the level of public awareness, and the diversity of potential invaders and native aquatic and riparian biota. High biodiversity in the Eastern United States, coupled with relatively few aquatic specialists, reduces the probability of an invasive taxon being recognized as nonnative. For some taxa (e.g., crayfish), complete ranges of many native species are not known, further compounding the difficulty of recognizing invasions. Both information transfer experts and field biologists need guidance in identifying organisms that should be the foci of their efforts. Forest Service R&D can assist by doing the following:

1. Developing regional lists of potential invasive aquatic and riparian species.

2. Creating or contributing to taxonomic guides and voucher specimen collections.

3. Collaborating to train NFS biologists and other partners to detect invasive species.

4. Developing a sentinel strategy, including identifying sites where introductions are most likely to occur and developing rapid survey protocols for monitoring these sites, increasing the likelihood of detecting new invasions while populations are small, localized, and still vulnerable to eradication.

Eradication of aquatic invasive species is often difficult or impossible over large areas but may be successful as part of a targeted rapid response to incipient aquatic invasions. Active control measures for aquatic invasive species have been criticized as costly, ineffective, and damaging to some native plants and animals. Research to better determine impacts of existing control measures, to develop mitigation strategies, and to evaluate alternative control measures may be of great interest to Forest Service R&D customers. This research area is one of potential strength for Forest Service R&D, in part because NFS biologists and Forest Service engineers (e.g., in the San Dimas and Missoula Technology and Development Centers) can participate in developing and implementing experimental control measures over long periods.

Passive control methods associated with maintaining and restoring natural hydrologic and thermal regimes may be far more effective and efficient than active control with respect to many invaders, but research is needed to better understand conditions under which passive control is effective. Invasive fish, aquatic invertebrates, and riparian plants often establish and thrive in altered or degraded habitats. Research focused on natural processes constraining the distribution of invasive species at local and regional scales (e.g., predictive models above) could lead to more efficient control measures.

Management and Mitigation

In most cases, attention to invasive aquatic and riparian species in the Forest Service has been inadequate for developing effective management options. Lack of understanding about the full range of ecological effects of specific invasive aquatic and riparian species on native plants and animals limits abilities to develop effective, science-based management options. Forest Service R&D could play an important role in providing basic ecological knowledge about invasive species and their effects.

Research is needed to better understand, manage, and mitigate effects of invaders across terrestrial-aquatic boundaries. Forest Service R&D results illustrate that nonnative fish introductions in high-elevation lakes can lead to food web effects that influence terrestrial wildlife. Similarly, terrestrial invaders can influence aquatic communities. Nonnative feral pigs alter stream invertebrate and microbial communities and increase pathogen levels (Kaller and Kelso 2006) and stream nitrate concentrations (Singer et al. 1984). Invasive riparian plants pose substantial threats to native aquatic species and may dramatically alter ecosystem functioning (Richardson et al. 2007), but research is just beginning in this area. Despite their potential threat, invasive riparian plants have received relatively little attention.

Restoration and Rehabilitation

Large amounts of money are spent nationally to restore aquatic habitats, typically with the ultimate goal of recovering or reestablishing native aquatic fauna. In some cases, invasive species have immediately colonized and thrived in the restored habitat, rendering the restoration unsuccessful for conservation purposes. To better prioritize funds for habitat restoration, research is needed to predict the circumstances under which restored habitats are likely to be invaded. Because of Forest Service involvement in restoration, this role is a logical one for Forest Service R&D.

Application and Communication

Responding quickly to new invasive species increases the probability of eradication and can minimize negative ecosystem effects. The Forest Service does not have a coordinated strategy to identify, rapidly respond to, and prioritize invasive aquatic and riparian species threats and research needs at national or regional levels. "Outbreaks" are typically managed at the local level without the benefit of regionwide coordination and technical information transfer. In addition to needing intra-agency coordination, the Forest Service needs mechanisms in place for rapid communication with external scientists. NFS or Forest Service R&D representation on Aquatic Nuisance Species Task Force regional panels or research prioritization committees may be an effective means for meeting this need, while also serving as a means for Forest Service scientists to become more familiar with regional invasive species issues and the people working on them. Forest Service R&D also needs to improve communication with regulators and policymakers involved in invasive species issues.

Because of close interaction with a national network of NFS biologists and botanists, Forest Service research scientists are well positioned to both obtain information from and provide research results to the field. Formalizing these relationships with regard to invasive species information may encourage and facilitate such communication.

Top Research Needs

The following list enumerates some important general research needs, but prioritization of specific research needs, although important, will require a more thorough and inclusive approach than our timeline has allowed. If Forest Service R&D increases emphasis on aquatic and riparian invasive species research, the first steps should be a thorough inventory of institutional capacity to conduct such research, followed by a more systematic regional or national prioritization of aquatic invasive species research needs. The prioritization can be based on existing knowledge but should follow a formal process. Participants can be expected to include scientists from a broad range of disciplines (e.g., hydrology and geomorphology as well as stream ecology and fish biology) and NFS managers and professionals knowledgeable about aquatic and riparian invasive species issues.

Within the stated context, we offer the following eight top general research needs.

1. Develop new prediction and ecological risk assessment tools essential for helping decisionmakers prioritize which invasive species to address, what actions to take, and where to take them. In many cases, useful data exist, emerging statistical approaches offer greater power than ever before, and decision support and prioritization frameworks are available for consistent analysis and effective communication. Despite these available resources, developing ecological risk assessment tools will require a substantial investment in new data and models to predict probable invasions, species interactions, and ecological outcomes. Initial modeling efforts can focus on potential and established invaders that appear to pose the most serious risks. Effective prediction and prioritization must be conducted in the context of large-scale influences. Fire, climate change, and changing forest community composition are clearly important cross-cutting issues because changing environments will alter the constraints on species distributions.

2. Contribute to building and maintaining state-of-the-art, centralized data repositories. This action is critical for documenting species spread and for risk assessment and model development. Forest Service R&D can collaborate with the NFS and with other agencies already managing aquatic and riparian invasive species data. Examples of existing national databases are those maintained by the USGS Nonindigenous Aquatic Species program in Gainesville, FL, and by the USGS National Institute of Invasive Species in Fort Collins, CO. Biologists and fire personnel, however, often require distribution data at finer spatial scales than are now available to map invasive species occurrences at district or more local levels.

3. Increase awareness and investigation of the interaction between global climate change and aquatic invasive species. Conduct research on linkages between region-specific climate projections and invasion risk and on synergistic effects of climate change and nonnative species on native communities.

4. Improve understanding of ecological, physical, and biological factors facilitating and inhibiting invasions. Encourage research to move beyond *species-habitat relationships* toward investigations of *species interactions* (which is key to understanding effects of invasive species) as influenced by habitat and disturbance. Examine effects of invasions on ecosystem functions.

5. Increase multiscale research to better understand and model the hierarchy of controls on invasions. For example, an effective research approach to large-scale invasions may be to explore invasion patterns and associations at a variety of scales to develop hypotheses regarding controls, conduct mechanistic research at appropriate scales, and then reaggregate results for prediction across scales.

6. Develop more effective prevention/eradication/control measures, and use risk assessment tools for weighing the potential benefits versus deleterious effects on native species.

7. Enhance the roles of social sciences and economics in aquatic invasive species research. For example, understanding conflicting public values is important in developing valid risk analyses and successful control strategies for aquatic and riparian invasive species. Evaluating the efficacy of different outreach strategies can identify tools that increase public motivation and, thus, compliance with preventative measures. Accounting for the full costs of species invasions will be instrumental for informing the public and policymakers of potential societal effects from nonnative species and, thus, for adopting effective prevention and control strategies.

8. Improve communication of invasive species science among scientists, NFS managers, policymakers, and the public. Although not a research need *per se*, a need exists to bring Forest Service scientific expertise to bear on issues of policy and regulation aimed at preventing future aquatic and riparian invasions. Establish mechanisms for rapid participation *as an agency* to provide science-based input on questions of transporting species across national and natural boundaries.

Structuring Forest Service R&D for Effective Aquatic Invasive Species Research

Although issues surrounding aquatic and riparian invasions will certainly intensify, predicting specific issues is impossible. Thus, Forest Service R&D can be best prepared by maintaining broad expertise within a structure flexible enough to respond rapidly to new threats. The Panel noted a need to maintain broad taxonomic expertise, which Forest Service R&D could improve on. Our representation (in terms of numbers and distribution of positions) in fish ecology is strong in the West and somewhat weaker in the East, and our expertise in herpetology is scattered. Expertise in riparian and aquatic plants, mussels, crayfish, and aquatic insects is scarce in Forest Service R&D. The Forest Service can provide ecological and landscape scale research on invasive aquatic diseases and parasites, but collaboration with research organizations operating established disease laboratories would maximize efficiency. External collaborations will, of course, remain essential to our overall effectiveness in invasive species research.

Other disciplines are also necessary for a successful invasive species research program. Many aspects of aquatic ecological research, including invasion research, require expertise in genetics. Genetics work is currently accomplished primarily through external collaboration, but Forest Service R&D may consider the cost-effectiveness of increasing capabilities internally, as a national resource. Expertise in geographic information systems (especially with regard to stream networks), spatial analysis, epidemiological modeling, and risk assessment varies by research station, but is essential to developing predictive models and integrating invasive species data management for an effective aquatic invasive species research program.

Critical to quickly and cost-effectively responding to invasion issues is not overcentralizing expertise. Invasion issues are typically region specific; thus, addressing them depends on maintaining regional understanding and awareness. Because the aquatic research program is small and invasive species issues are numerous and often region specific, duplication of aquatic research effort is not a problem and likely will not become a problem within Forest Service R&D in the near future.

Having argued for dispersed expertise, we acknowledge that some skills may be in common demand nationally. For example, the suite of analytical and predictive tools for risk analysis and prediction of species habitat and occurrence might be collaboratively developed and maintained in a "center of excellence" but fed by data and research from all regions. For common species groups or guilds, developing common approaches could be powerful. Sharing knowledge and data for species that are native in one region and invasive in another could facilitate understanding of the primary constraints and development of the needed predictive models. In many cases, broad collaboration provides the foundation for understanding that may otherwise be impossible.

Scientists initiate most cross-station research efforts. Future collaboration could be fostered through national or multiregion panels, composed of Forest Service R&D scientists, NFS personnel, and other stakeholders, identifying important issues and then funding relevant research. Key challenges of predicting, preventing, and controlling invasive species may be best met by combining multistation teams of scientists who have local, spatially explicit knowledge of conditions and key processes. Teams could focus on (1) identifying and studying taxa that are important over large areas and (2) refining risk assessment models. These large-scale efforts would identify taxon- and context-specific needs for research on combinations of potentially invasive taxa and ecologically important resources.

A byproduct of national teams would be better communication among Forest Service aquatic scientists. Identifying particular expertise in aquatic science is difficult within the Forest Service. Mechanisms (e.g., a Web-accessible database) for locating Forest Service scientists with various skills related to invasive species would facilitate communication; however, we strongly agree with the Panel's caution about increasing the reporting burden on scientists.

Finally, we identified one of our strengths as the ability to do long-term research. This strength could be greatly improved by lengthening the research funding cycle or creating better mechanisms for funding long-term research.

Conclusions

Uncertainty will always be a major feature of invasive species science. Confronting uncertainty is important in both management and research and in communicating with stakeholders. Principles articulated by Ludwig et al. (1993) suggest that managers should favor decisions that are robust to uncertainty (i.e., the outcome is likely to be favorable regardless of the result). If that is not possible, it is still important to hedge (use a mix of strategies), favor reversible decisions, and intentionally probe ecosystems to learn through adaptive management (try some risky things for the sake of learning) (e.g., Fausch et al. 2006). The Forest Service has long advocated adaptive management but has seldom implemented it with true management experiments designed for learning (e.g., Bisson et al. 2003). Invasive species issues provide an opportunity for the marriage of research and management in the Forest Service that could be extremely important to future management.

Our society highly values aquatic recreation, clean water, and freshwater biodiversity. The future of all three components depends on strong research programs to address the ever-increasing threat of invasive species that can profoundly alter our waters and riparian areas. Thus, we deem it essential for the Forest Service to commit to a research program and infrastructure that will effectively address invasive aquatic and riparian species issues.

Acknowledgments

The authors thank Nick Schmal, Glen Contreras, Kate Dwire, Frank McCormick, Rich MacKenzie, Hart Welsh (Forest Service) and Jason Dunham (U.S. Geological Survey) for insightful discussions about the Forest Service role in aquatic invasive species research. In addition, Lindsay Chadderton (The Nature Conservancy), Steve Lindley (National Oceanic and Atmospheric Administration), Glen Contreras, Doug Ryan, Nick Schmal, Frank McCormick, Dede Olson, Mel Warren, Mike Young (Forest Service), and three anonymous reviewers provided helpful reviews of an earlier draft. Tami Lowry (Forest Service) formatted the document.

Literature Cited

Bisson, P.A.; Rieman, B.E.; Luce, C., et al. 2003. Fire and aquatic ecosystems of the Western USA: current knowledge and key questions. Forest Ecology and Management. 178(1–2): 213–229.

Brown, L.R.; Moyle, P.B. 1997. Invading species in the Eel River, California: successes, failures, and relationships with resident species. Environmental Biology of Fishes. 49: 271–291.

Fausch, K.D.; Rieman, R.E.; Young, M.K.; Dunham, J.B. 2006. Strategies for conserving native salmonid populations at risk from nonnative fish invasions: tradeoffs in using barriers to upstream movement. Gen. Tech. Rep. RMRS-GTR-174. Fort Collins, CO: U.S. Department of Agriculture, Forest Service, Rocky Mountain Research Station. 44 p.

Great Lakes Regional Collaboration (GLRC). 2005. Great Lakes Regional Collaboration Strategy. http://www.glrc.us/strategy.html. (5 March 2007).

Harvey, B.C.; White, J.L.; Nakamoto, R.J. 2004. An emergent multiple predator effect may enhance biotic resistance in a stream fish assemblage. Ecology. 85: 127–133.

Kaller, M.D.; Kelso, W.E. 2006. Swine activity alters invertebrate and microbial communities in a Coastal Plain watershed. American Midland Naturalist. 156: 163–177.

Lee, D.C.; Sedell, J.R.; Rieman, B.E., et al. 1997. Broadscale assessment of aquatic species and habitats. In: Quigley, T.M.; Arbelbide, S.J., tech. eds. An assessment of ecosystem components in the interior Columbia basin and portions of the Klamath and Great Basins. Gen. Tech. Rep. PNW-GTR-405 (Vol. 3). Portland, OR: U.S. Department of Agriculture, Forest Service, Pacific Northwest Research Station: 1057–1496.

Lodge, D.M.; Williams, S.; MacIsaac, H.J., et al. 2006. Biological invasions: recommendations for U.S. policy and management. Ecological Applications. 16: 2035–2054.

Ludwig, D.; Hilborn, R.; Walters, C. 1993. Uncertainty, resource exploitation, and conservation: lessons from history. Science. 260: 17–36.

Master, L.L.; Stein, B.A.; Kutner, L.S.; Hammerson, G.A. 2000. Vanishing assets—conservation status of U.S. species. In: Stein, B.A.; Kutner, L.S.; Adams, J.S., eds. Precious heritage: the status of biodiversity in the United States. New York: Oxford University Press: 93–118.

National Invasive Species Council (NISC) and Aquatic Nuisance Species Task Force (ANSTF). 2007. Training and implementation guide for pathway definition, risk analysis and risk prioritization. http://www.anstaskforce.gov/Documents/Pathways_Training_and_Implementation_Guide_Jan_2007.pdf. (23 March).

National Research Council. 2004. Atlantic salmon in Maine. Washington, DC: National Academy Press. 304 p.

Nislow, K.H.; Magilligan, F.J.; Fassnacht, H., et al. 2002. Effects of hydrologic alteration on flood regime of natural floodplain communities in the Upper Connecticut River. Journal of the American Water Resources Association. 38(6): 1533–1548.

Raffia, K.F.; Bonello, P.; Cameron, R.S., et al. 2006. USDA Forest Service invasive species research and development panel. Report submitted to the Forest Service, Washington, DC. 16 p.

Ricciardi, A. 2006. Patterns of invasion in the Laurentian Great Lakes in relation to changes in vector activity. Diversity and Distributions. 12: 425–433.

Richardson, D.M.; Holmes, P.M.; Else, K.J., et al. 2007. Riparian vegetation: degradation, alien plant invasions, and restoration prospects. Diversity and Distributions. 13: 126–139.

Rinne, J.N. 1996. The effects of introduced fishes on native fishes: Arizona, Southwestern United States. In: Philipp, D.P., ed. Protection of aquatic diversity: proceedings of the First World Fisheries Congress. New Deli, India: Oxford & IBH Publishing: 149–159.

Rinne, J.N.; Medina A.L. 1996. Implications of multiple use management strategies on native southwestern (USA) fishes. In: Meyer, R.M., ed. Fisheries resource utilization and policy: proceedings of the First World Fisheries Congress. New Deli, India: Oxford & IBH Publishing: 110–123.

Singer, F.J.; Swank, W.T.; Clebsch, E.E.C. 1984. Effects of wild pig Sus scrofa rooting in a deciduous forest. Journal of Wildlife Management. 48: 464–473.

U.S. Department of Agriculture, Forest Service. 1995. Forest service manual, title 2600, wildlife, fish and sensitive plant habitat management, amendment no. 2600-95-7, section 2670.22. Washington, DC: U.S. Department of Agriculture, Forest Service.

U.S. Federal Aquatic Nuisance Species Task Force (USFAN-STF). [N.d.] ANS task force. http://www.anstaskforce.gov. (28 March 2007).

U.S. Geological Survey (USGS). 2008. NAS—nonindigenous aquatic species. http://nas.er.usgs.gov/queries/default.asp. (7 August 2000).

Wilcove, D.S.; Rothstein, D.; Dubow, J., et al. 2000. Leading threats to biodiversity: what's imperiling U.S. species? In: Stein, B.A.; Kutner, L.S.; Adams, J.S., eds. Precious heritage: the status of biodiversity in the United States. New York: Oxford University Press: 239–254.

Invasive Soil Organisms and Their Effects on Belowground Processes

Erik Lilleskov[1], Mac A. Callaham, Jr.[2], Richard Pouyat[3],
Jane E. Smith[4], Michael Castellano[5], Grizelle González[6],
D. Jean Lodge[7], Rachel Arango[8], and Frederick Green[9]

Abstract

Invasive species have a wide range of effects on soils and their inhabitants. By altering soils, through their direct effects on native soil organisms (including plants), and by their interaction with the aboveground environment, invasive soil organisms can have dramatic effects on the environment, the economy and human health. The most widely recognized effects include damage to human health and economies, such as that caused by invasive fire ants and termites. Many other soil invasive species, however, have pervasive but poorly understood effects on terrestrial ecosystems. These species include the following:

1. Invasive plants and their symbionts (e.g., *Falcataria* in Hawaii).

2. Herbivores (e.g., root-feeding weevils).

3. Ecosystem engineers (e.g., earthworms).

4. Keystone species (e.g., terrestrial planaria).

In addition, aboveground invasive species, notably herbivores and pathogens, can have major indirect effects on belowground processes by altering nutrient cycles, plant health, productivity and carbon (C) allocation patterns, demography, and community composition and function.

Given the diversity of invasive soil organisms, there is a need for Forest Service Research and Development (R&D) to develop a prioritized list of invaders and research topics to help guide research and identify research gaps. Large gaps exist in our knowledge of the identity, distribution, abundance, and effects of most invasive soil organisms. Organisms with uncertain but potentially large ecosystem effects (e.g., invasive planaria) deserve more attention. In addition, we perceive several areas emerging as important research topics for Forest Service R&D. These topics include the widespread increase in propagule pressure of soil invasive species in urban areas and in the wildland-urban interface, the potential for additive and synergistic effects of suites of soil invasive species, the feedbacks between invasive species and soil microbial communities, and the interactions of soil invasive species with global change.

All stages of management of soil invasive species are critical, and Forest Service R&D is poised to play a leadership role. In the prediction and prevention area, we are in need of a more coordinated effort. Forest Service R&D has the expertise to inform the U.S. Department of Agriculture (USDA), Animal and Plant Health Inspection Service (APHIS) and other organizations about gaps in their programs for excluding or limiting dispersal of soil invasive species, but, at present, no comprehensive program exists to generate such information. Some work is being done on biogeographic models of invasive distribution that could inform prediction and prevention efforts. In the detection and eradication, management and mitigation, and restoration and rehabilitation areas, we have scientists directly addressing major soil invasive species issues, including

[1] Research Ecologist, Forest Service, Northern Research Station, Forestry Sciences Laboratory, 410 MacInnes Dr., Houghton, MI 49931.

[2] Research Ecologist, Forest Service, Southern Research Station, Forestry Sciences Laboratory, 320 Green St., Athens, GA 30602-2044.

[3] Bioclimatologist, Forest Service, Environmental Science Research Staff, Rosslyn Plaza, Building C, 1601 N. Kent St., 4th Floor, Arlington, VA 22209.

[4] Botanist, Forest Service, Pacific Northwest Research Station, 3200 SW Jefferson Way, Corvallis, OR 97331.

[5] Research Forester, Forest Service, Northern Research Station, c/o Pacific Northwest Research Station, 3200 SW Jefferson Way, Corvallis, OR 97331.

[6] Research Ecologist and Director, Forest Service, Sabana Field Research Station, International Institute of Tropical Forestry, Jardín Botánico Sur, 1201 Calle Ceiba, San Juan, PR, 00926-1119.

[7] Botanist, Forest Service, Northern Research Station, P.O. Box 1377, Luquillo, PR 00773-1377.

[8] Entomologist, Forest Service, Forest Products Laboratory, One Gifford Pinchot Dr., Madison, WI 53726-2398.

[9] Microbiologist, Forest Service, Forest Products Laboratory, One Gifford Pinchot Dr., Madison, WI 53726-2398.

effects and control of invasive termites; belowground effects of invasive plant species; interactions of invasive plants with soil microbial and fungal communities; effects and management of invasive earthworms; diversity and effects of urban soil invasive species; diversity, distribution, and effects of root-feeding weevils; and biogeography of invasive soil macroinvertebrates.

The Forest Service has strengths that permit us to directly address these problems, including a network of scientists investigating soil invasive species. Some gaps do exist in our expertise, however, most notably in taxonomy of soil organisms. These gaps should be addressed via either new hires or collaboration with non-Forest Service scientists.

We need to do a better job of communicating the diversity of Forest Service research in this area, both internally and externally. Increased opportunities for communication among Forest Service scientists working in this area would facilitate our efforts, and expansion of the invasive species Web site to include a section on soil invasive species would improve communication of our results. The Forest Service should host regular national meetings on soil invasive species to link Forest Service and other scientists and managers.

The continued erosion of the Forest Service research budget jeopardizes all these efforts. Long-term efforts in managing the effects of invasive species will require significantly expanded investments in Forest Service R&D. Maintenance of the status quo (or, worse, continued budget erosion) will contribute to the Nation's inability to cope with potential ecological disasters, such as the chestnut blight epidemic or, more recently, the emerald ash borer invasion, which have transformed, or are in the process of transforming, entire forest ecosystems.

Introduction

Soils are the foundation of productive ecosystems, providing a matrix within which plant roots provide support and forage for nutrients and water. Their properties derive from complex interactions of physical, chemical, and biological processes that drive the cycling and storage of carbon and nutrients. The biological processes are carried out by a highly diverse and complex array of plants, microorganisms, fungi, invertebrates, and vertebrates. This biodiversity is essential for the production of ecosystem goods and services, such as timber and nontimber

forest products; as a source of new pharmaceuticals and other products derived from plants, bacteria, and fungi; for protection of clean air and water; for mitigation of changing atmospheric chemistry via carbon sequestration in soils, in the plants sustained by them, and in forest products; for protection of habitat for game and nongame wildlife alike; and for provision of recreational opportunities for the millions of people who enjoy the Nation's forests every year.

These goods and services are very sensitive to the biotic communities that control them; therefore, invasive species have a large potential effect on them. Plant and animal species can alter nutrient and water cycling, rates of decomposition and storage of soil C, soil structure and fertility, tree growth and mortality, and a host of other properties. For example, the introduction of a single "ecosystem engineer," such as an invasive earthworm species, has the potential to completely alter the chemical and structural properties of soils. In addition to having direct effects on soils, soil organisms can affect other species that live in or use the soil (e.g., via root herbivory or disease). The introduction of predators can alter the invertebrate communities in soils, with potential effects on soil processes and on forest food webs dependent on those soil organisms. The most obvious effects of introduced soil organisms are on human health and economies (e.g., introduced fire ants and Formosan subterranean termites), but the other effects described previously are likely to have more significant environmental and economic consequences that are at present only poorly quantified.

In this vision paper, we describe what we see as the most pressing issues surrounding the question of invasive species in soils, highlighting both what we already know and what we consider to be important knowledge gaps. We address specific taxonomic/functional groups, emerging broad issues, and issues related to the specific steps involved in responding to invasive species. We also highlight important strengths and weaknesses of Forest Service R&D in our capacity to address soil invasive species issues.

Effects of Taxonomic and Functional Groups on Belowground Processes

Taxonomic and functional groups differ in their mode of effect, so we present here a brief summary of the direct effects of the major groups of invasive soil organisms and the Forest Service R&D efforts related to each group.

Plants

Invasive plant effects on ecosystems can be substantial. They have been summarized in extensive reviews of this topic elsewhere (e.g., Ehrenfeld 2003, Reichard and Hamilton 1997, Stohlgren et al. 2004) and are being addressed in another vision paper, so we mention them only briefly in the context of belowground effects. Plant belowground effects are expected to be greatest when a new functional group enters a region; for example phreatophytic *Tamarix* in riparian zones leading to degradation of riparian zones, nitrogen fixing *Myrica faya* and *Falcataria* in Hawaii leading to loss of native biological diversity (Vitousek and Walker 1989), and alien plants with traits leading to enhanced fire regimes with devastating effects on native organisms and dramatic alterations of ecosystem processes (Brooks et al. 2004). Forest Service R&D has extensive efforts addressing invasive plant species effects on belowground processes (see "The Role of Forest Service in Nonnative Invasive Plant Research," chapter 3, Sieg et al. 2010 in this document).

Aboveground Diseases and Herbivores: Indirect Effects

The pathogen and insect vision papers will be addressing these topics in detail, so we address these only lightly here. Indirect effects of aboveground herbivores occur via alteration of aboveground conditions or processes in ways that have belowground effects. These effects include those of aboveground herbivores on rates of plant growth, mortality, and litterfall. These changes can significantly alter nutrient cycles and disturbance regimes (e.g., windthrow, fire), with consequences within the affected site and with downstream effects on hydrology and stream chemistry (Ellison et al. 2005). Forest Service R&D has extensive research in these areas (see "Forest Service R&D—Invasive Insects: Visions for the Future" and "Invasive forest pathogens: Summary of issues, critical needs, and future goals for USDA Forest Service Research and Development," chapter 2, Klopfenstein et al. 2010 in this document).

Similarly, changes in belowground communities have the potential to affect aboveground herbivores and diseases by a variety of pathways (Scheu 2001).

Belowground Diseases and Herbivores

In addition to the indirect effects of aboveground diseases and herbivores, there are direct belowground effects of root diseases and herbivores. The disease effects are exemplified by *Phytophthora lateralis* root rot on Port Orford cedar. Forest Service pathologists have been involved in investigating this disease for decades (e.g., Greenup 1998, Zobel et al. 1982). An emerging area of interest is the potential for soil microbial communities to structure the interactions between native and invasive plant species (Klironomos 2002). This topic will be addressed in more detail below (see Key Issues).

Root-feeding herbivores can also have large effects on plants and ecosystems, although our understanding of the importance of this phenomenon is hindered by the paucity of studies of root herbivory compared with aboveground herbivory (Blossey and Hunt-Joshi 2003, Hunter 2001, Masters 2004). Root feeders come from a variety of taxonomic groups but are especially well represented in the Coleoptera, Lepidoptera, and Nematoda. Effects of these herbivores can be seen in altered root demography (Stevens et al. 2002, Wells et al. 2002), plant growth and seed production, root:shoot ratios, nutrient status (Masters 2004), multitrophic interactions, (Masters 2004, Van der Putten et al. 2001), and plant community structure (Gange and Brown 2002). One indication of the importance of root-feeding herbivores is found in the literature on classical biological control. Approximately 65 percent (20 of 31) of intentionally released Coleoptera in the Chrysomelidae (8 of 12), Curculionidae (9 of 14), Cerambycidae (1 of 3), and Buprestidae (2 of 2) have contributed to control of invasive plant species (Blossey and Hunt-Joshi 2003).

Therefore, the widespread abundance of a variety of invasive root herbivores is likely to have far-reaching ecological effects in natural forests. Although only a small proportion of these species are introduced, they are often widely distributed and abundant and can locally outnumber native root herbivores (Pinski et al. 2005a, 2005b). The Forest Service Northern Research Station (NRS) (Mattson, Friend, Lilleskov) and collaborators (K. Raffa, D. Coyle) are leading this research effort in the northern region. Biogeographic patterns of their distribution have also been investigated, pointing to northern peaks in abundance of introduced root-feeding weevils (Lilleskov et al. 2008).

Predators/Keystone Species

Introduced predators have been described as keystone species when their actions have significant top-down effects on food web structure, community composition, and ecosystem processes, although some question the usefulness of this concept (e.g., Kotliar 2000). Groups of ecologically important introduced predators affecting soil communities include ground beetles, centipedes, and planaria. Invasive terrestrial planaria could have major effects on ecosystem processes, because many of these species of flatworms are predators of earthworms, attacking and killing individuals more than 10 times their size (Ducey et al. 1999). One species, *Bipalium adventitium*, has been found in at least eight States (Ogren and Kawakatsu 1998) and preys on a diversity of earthworm species (Ducey et al. 1999, Fiore et al. 2004, Zaborski 2002). They are currently concentrated in human-altered habitats such as lawns and gardens, and it is not known whether this constrained habitat range is due to environmental limitations or slow dispersal into native habitats (Ducey and Noce 1998). Another invasive planarian has been shown to significantly reduce the abundance of earthworms, with apparent cascading effects on other earthworm predators (Boag and Yeates 2001). Thus, it may function in North America as a broad host range biological control agent with negative effects on native as well as introduced earthworm species. The ecological consequences will depend on the rate of spread and efficacy of these predators but could have significant effects on earthworm-mediated processes in both agroecosystems and forests. Other introduced predators of concern are Carabidae (ground beetles), which can become numerically dominant in certain habitats. Their effects on communities and ecosystems are unclear, beyond possible reductions in native ground beetle abundance (Spence and Spence 1988). Introduced centipedes could also have significant effects via predation and competition with native predators, but relatively little work has been done on their diversity, distribution, and effects (Hickerson et al. 2005). Forest Service R&D active research on keystone species in soils is limited to a recent analysis of biogeography of introduced ground beetles (Lilleskov et al. 2008).

Ecosystem Engineers

Many invertebrates have been characterized as ecosystem engineers (Jones et al. 1994) because of their ability to alter ecosystem properties and processes disproportionately to their biomass or food consumption. Some of the best examples of invasive ecosystem engineers include earthworms, ants, and termites.

Earthworms

By consuming soil organic matter and mixing soils, earthworms have a dramatic effect on soils and the ecosystems they support. Two distinct science problems associated with invasive earthworms in North America are (1) invasions north of the Wisconsinan glacial boundary where no native earthworm species reside and (2) invasions south of this boundary where invasive species may interact with native earthworms. In some regions of North America, invasive earthworm species often have greater species richness and abundance than natives. In fact, in much of the previously glaciated regions of North America, the earthworm fauna is composed exclusively of invasive species (Lilleskov et al. 2008, Reynolds and Wetzel 2004).

Forest Service scientists in the International Institute of Tropical Forestry (IITF) (González) and Southern Research Station (SRS) (Callaham) have been working on problems associated with invasive earthworms for about 10 years each in the Caribbean tropics and the Southern United States, respectively (Callaham and Blair 1999, Callaham et al. 2003, Callaham et al. 2006a, Callaham et al. 2006b, González 2002, González 2006, González and Seastedt 2001, González and Zou 1999, González et al. 1996, González et al. 1999, González et al. 2003, González et al. 2006a, González et al. 2006b, González et al. 2007a, González et al. 2008, González et al. 2007b, Huang et al. 2006, Yiqing and González 2008, Zou and González 1997). They have been involved with an international group of researchers who are concerned with earthworm invasions, and both contributed to a special issue of the journal *Biological Invasions,* which focused on the topic (Callaham et al. 2006a, González et al. 2006, Hendrix et al. 2006). In addition, IITF (González) hosted the second Latin American Symposium of Earthworm Ecology and Taxonomy. The peer-reviewed proceedings of the meeting were published as a special issue in the *Caribbean Journal of Science* (González 2006).

In regions where only nonnative earthworms are present, we are seeing a fundamental shift in soil properties from unmixed soils with distinct organic horizons (mor) to well-mixed soils without organic horizons (mull). These changes have profound effects on soil properties and processes that ramify throughout ecosystems, including emerging problems such as siltation of water sources resulting from increased soil erosion and threats to endangered herbaceous plants (e.g., Bohlen et al. 2004a, Bohlen et al. 2004b, Gundale 2002, Hale et al. 2005, Hale et al. 2006, Steinberg et al. 1997). Reductions in arbuscular mycorrhizal fungal abundance and colonization in the presence of invasive earthworms have already been demonstrated (Lawrence et al. 2003) and are implicated as one of the factors in sugar maple decline and loss of understory species (Bohlen et al. 2004b, Frelich et al. 2006, Holdsworth et al. 2007). NRS scientists are studying ecosystem effects of earthworm invasions in urban ecosystems (Pouyat) and north of the glacial boundary (Lilleskov, Kolka, Swanston) to determine how forest management must be adapted to this state-shift in soil properties.

Termites

Termites have major economic and ecological effects via their consumption of wood and other organic matter, via bioturbation, and via emissions of the greenhouse gas methane. Given the large economic effects, most efforts on invasive termites have focused on their effects on manmade structures; however, the effects of invasive termites on forest ecosystems are less studied and in need of more attention. Termites have been considered ecosystem engineers (Dangerfield et al. 1998) because they have the potential to significantly alter rates of bioturbation of soils, carbon cycling, trace gas emissions (Sugimoto et al. 1998, Wheeler et al. 1996), and other ecosystem properties. The most widespread and economically costly invasive termite in the United States is the Formosan subterranean termite (*Coptotermes formosanus*). This species can form supercolonies and hollow out live trees for its nests. The potential economic and ecological effects of these changes in forest ecosystems are extensive and are being investigated by Forest Service scientists in the SRS (T. Wagner).

Ants

Large numbers of invasive ant species exist, especially in the Southern United States (Lilleskov et al. 2008). Invasive ants affect the economy and human health (e.g., fire ants), ecological processes such as pollinator interactions, seed dispersal, and native biodiversity (Christian 2001, Holway et al. 2002). Invasive ants such as *Linepithema humila* and *Solenopsis* sp. have reduced parasite loads in their introduced ranges and are likely candidates for introduction of biocontrol agents (Chen 2004, Feener 2000), because their lower genetic diversity may make them particularly sensitive to this form of management. The potential exists for invasive wood ants to have significant effects in northern forests, such as populations intentionally established in eastern Canada in misguided biocontrol efforts (Jurgensen et al. 2005). In addition, invasive European fire ants are spreading in the Northeastern United States (Groden

et al. 2005) and could have significant effects on human and ecosystem health as they aggressively attack organisms near their nests. Beyond a biogeographic review of invasive ants (Lilleskov et al. 2008), we are not aware of any Forest Service R&D active projects on invasive ants.

Other Invertebrates

Millipedes, isopods, and gastropods can all be numerically important invasive species in some ecosystems. Their effects on ecosystem processes have only been superficially examined. Detritivory, fungivory, herbivory, and predation by invasive gastropods could have significant effects (e.g., Tupen and Roth 2001). Invasive isopods may have significant effects via direct feeding on fungal sporocarps. Like earthworms, the isopod fauna north of the glacial maximum is dominated by introduced Eurasian species (Jass and Klausmeier 2000), but little investigation into their effects and potential for spread into undisturbed ecosystems has been carried out. The native vs. introduced status of many taxa, especially meso- to microinvertebrates (e.g., mites and collembola), which are numerically dominant in northern forest soils, cannot be assessed because of incomplete knowledge of taxonomy and biogeography.

Vertebrates

Invasive vertebrates with indirect effects are covered in another vision paper. The invasive with the greatest direct effects on forest soils is the feral pig (*Sus scrofa*). This species is of great concern in both the continental United States and Hawaii (Singer 1981) via its role in uprooting plants, bioturbation, and facilitation of invasive plant species.

Plant Symbionts: *Mycorrhizal fungi*

Recognition of the need to consider the possible negative effects of introduction of nonnative mycorrhizal fungi is growing (Schwartz et al. 2006). Fungi from different parts of the world have been introduced with plants or in soil (e.g., in Puerto Rico, Florida, and California, along with nonnative *Pinus, Eucalyptus, Allocasuarina,* or *Casuarina*). *Eucalyptus* is widespread in California and has slowly escaped from its original plantings to occupy nearby habitat, leading to the replacement of native ectomycorrhizal fungi with the introduced species associates (Castellano 2008). One area in need of investigation is the potential for invasive species to cause the extirpation of rare or threatened ectomycorrhizal fungi (e.g., in coastal California). Another area of concern is the redistribution of native ectomycorrhizal fungi via inoculation or plantings and the potential for these fungi to alter local genetic diversity. In addition, a real potential exists that importation of mycorrhizal inoculum could result in the introduction of diseases that could attack native fungi or plants, with serious but as yet unexplored consequences (Schwartz et al. 2006). Pacific Northwest Research Station (PNWRS) scientists (Castellano, Smith) are investigating some aspects of this problem. As the climate changes, we will see invasions of ectomycorrhizal species from the south, as evidenced from biogeographic studies (Mueller et al. 2007, Ortiz-Santana 2006). Biogeographic studies of ectomycorrhizal fungi and other root-associated fungi are of critical importance for invasive species predictions and are being conducted by Forest Service scientists in the NRS and PNWRS (Lodge 2001, Mueller et al. 2007, Ortiz-Santana 2006). Large-scale integrated approaches to characterizing these communities are essential to our ability to be able to define baseline biogeographic patterns and the effect of invasive species and other agents of environmental change (Lilleskov and Parrent 2007).

Role of Forest Service R&D at Different Stages of Invasion

Different strategies are appropriate at the various stages of invasion. In the following paragraphs we identify the key actions that are being taken or should be taken by Forest Service R&D at different stages of invasion.

Prediction and Prevention

Forest Service R&D can contribute to prediction and prevention of soil invasive species via a variety of actions, including characterizing invasion pathways and modeling efforts.

Characterizing Invasion Pathways

Understanding pathways of introduction, pools of potential invaders, and the risk of invasion for specific regions is critical to prevention and prediction of invasive species. For invasive species with large effects on belowground processes, the primary historic pathway of introduction was likely the transport of soils for ballast, planting medium, or other purposes. Prevention measures should continue to be coordinated with other agencies, such as APHIS, to intercept new species at ports of entry. Although many invasive soil organisms are not currently targeted by APHIS, current APHIS regulations severely limit the transport of soils, which presumably has greatly reduced the influx of invasive soil organisms. This presumption, however, should be tested; to test this presumption, we need better characterization of the diversity and distribution of invasive soil organisms already present and determination of the rate of new introductions. Very little of this type of work has been done for soil organisms (see, for example Larson and Langor 1982), although existing quarantine records could be analyzed for trends in interception of soil organisms.

Within the United States, restrictions on soil movement are in place only for specific quarantine areas (e.g., for fire ants), which do not cover many invasive soil organisms. Suggestions for control will depend on the taxonomic group. For example, in the case of earthworms, stricter guidelines for the bait and horticultural industries would be needed to contain invasions (Callaham et al. 2006a). For the fish-bait industry, these regulations would ideally eliminate the commercial availability of the more aggressive and generalist species (in terms of habitat requirements). For the horticultural industry, we suggest minimum guidelines that would ensure materials shipped from areas with invasive earthworms into areas not yet affected by those species be certified "worm-free" (this is not unprecedented for soil invertebrates, *c.f.* the imported fire ant). In the case of other groups (e.g., gastropods), although existing regulations limit their distribution, these species are readily available for sale on the Internet, and the regulations do not appear to be rigorously enforced (Tupen and Roth 2001).

Other invasive species with direct and indirect effects on soils, including plants, pathogens, and insects (herbivores, predators, and saprotrophs), are being vectored at high rates in other ways (intentional and unintentional plant transport, wood products, etc.) that are discussed more fully in their taxonomic treatments contained in this volume. These groups deserve the greatest attention in prediction and prevention efforts because of their high likelihood of both transport and negative effects on natural and managed ecosystems.

Modeling

A variety of modeling approaches are necessary to optimize our response to invasive species. Bioeconomic models define the best targets for response to invasive species (Leung et al. 2002). Ecological niche models define potential distribution of invasive species (Peterson and Vieglais 2001) and could be used to define the likely ranges and effect of potential soil invaders. Related risk models are currently in development for prediction of likely invasions of exotic earthworms and their effects on soil properties and processes. Such models need to be developed on a species-specific basis for different species of actual or potential soil invasive species. *Forest Service R&D should coordinate with other Government agencies to develop predictive models as one means of prioritizing risks for soil invasive species for prevention efforts.*

Detection and Eradication

Detection and eradication, both key components of efforts to respond to invasive species, each provide specific challenges to Forest Service R&D.

Detection

Reliable and effective detection of invasive soil organisms will require extensive monitoring and survey of sensitive habitats combined with the ability to recognize invasive species when they are encountered. Taxonomic resources (scientists and state-of-the-art identification facilities) must be available to rapidly determine the identity of novel organisms. In addition to facilitating the critical abilities of traditional taxonomists, we must avail ourselves of molecular identification tools when appropriate. A network of monitoring sites combined with a centralized facility to process the sampled material from a given taxonomic group would be one workable solution to the problem of invasive species detection. Whatever the approach, these efforts require first and foremost a more thorough understanding of the diversity and distribution of soil invasive species in North America and the ability to respond rapidly.

Eradication

Eradication efforts for soil organisms are challenging because of the difficulty of determining presence of cryptic species and because of the severe disruption of soils that might be involved. Appropriate methods and intensity of effort will depend on the taxonomic group. Taxa must be ranked in terms of probability of invasion and the costs of invasion vs.

eradication. Although the economic costs of some invasions are obvious (e.g., Formosan subterranean termite), the costs of other taxa are beginning to be appreciated but have not been fully quantified (e.g., invasive earthworms).

Approaches that attempt eradication or control at the stage of initial establishment can be potentially cost effective. *Knowing which introduced and naturalizing species to target for eradication is the greatest challenge in management of invasive species, because we have only limited ability to predict which subset of introduced species will result in ecosystem-modifying invasions. The modeling approaches discussed under prevention should be developed as a prioritization tool for determining appropriate responses.*

Direct Treatment Approaches. Chemical treatment is feasible when the distribution of populations of invasive soil organisms is well known and limited spatially (e.g., Arango and Green 2007). Research is critical in the areas of what compounds are effective, how and when they should be applied in a forestry or urban setting, what size of invasion can be treated effectively, which invasive species can be treated in this way, and what potential nontarget effects might result from such treatment.

Management and Mitigation

Efforts to manage and mitigate soil invasives must include a variety of approaches, including efforts to slow the spread of key invasives; characterization of environmental constraints and habitat, landscape, and risk modeling to better understand the areas most likely to be strongly affected by specific invasives; land management approaches to minimizing the impacts of invasives; and biological control of invasives when appropriate.

Slowing the Spread

A variety of approaches are available to either slow the spread or reduce the effects of invasive species. These approaches include comprehensive programs, such as Slow the Spread for gypsy moths, which combine education, biological control, trapping, and other approaches to reduce the rate of gypsy moth spread.

Formosan subterranean termites are prime candidates for such an effort, especially given their potential effect on forest ecosystems. Forest Service R&D has efforts aimed at controlling the damage caused by invasive termites. The Forest Service, Forest Products Laboratory (FPL), in collaboration with the Southern Regional Research Center, is working on new wood preservatives (Clausen et al. 2007, Lebow et al. 2006) and termite bait toxicants (Rojas et al. 2004) to prevent damage by native and Formosan subterranean termites. In certain instances, native subterranean termites (e.g., *Reticulitermes flavipes*) can be transported out of their endemic area, thus becoming invasive in a new ecological environment (Arango and Green 2007) (FPL, S. Lebow and F. Green; SRS, Wagner).

Forest Service work being done on termite control will be most effective as part of a coordinated effort to slow the spread of invasive termites (e.g., the USDA Agricultural Research Service (ARS) Operation Full Stop [Lax and Osbrink 2003]). In any effort of this sort, one of the critical questions to ask is, what are the costs and benefits of slowing the spread vs. not acting? Analyses of these sorts require collaboration among biologists, ecologists, and economists.

Characterization of Environmental Constraints and Habitat, Landscape, and Risk Modeling

To manage for invasive species, it is critical that we develop an understanding of basic life history and habitat tolerances of key species already introduced. This effort involves a comprehensive laboratory- and field-based approach to determine physical and chemical constraints on species distribution. Combined with landscape and risk modeling, this information is a key tool in developing appropriate management strategies for the key invasive species.

Land Management Approaches

Evidence indicates that certain types of land management can limit the encroachment of nonnative soil organisms. Maintenance of ecosystem-appropriate disturbance regimes appears to have promise as a means of limiting the spread of introduced soil organisms (Callaham et al. 2003, Callaham et al. 2006b). Research into the mechanisms behind this observation will lead to better understanding of invasion dynamics and, ultimately, to the development of management prescriptions that take invasive species into account.

Biological Control

Biological control is a key tool for control of invasive species and should be used when appropriate. Despite its potential, our ability to use biological control has limitations. One of the greatest limitations is potential serious nontarget effects of control organisms (Simberloff and Stiling 1996). The risks of these nontarget effects have to be carefully weighed vs. the potential

benefits of control. In addition, for certain invasive species, competing economic interests could limit the applicability of biological control. For example, biological control of the earthworm *Lumbricus terrestris* (the nightcrawler), even if ecologically viable, is unlikely to be a socially and economically acceptable option because of the importance of this species to the bait industry. We are aware of no Forest Service biocontrol programs addressing invasive soil organisms, although ARS collaborators are investigating the potential for biocontrol of the Formosan subterranean termite.

Restoration and Rehabilitation

Many soil invasive species are so completely naturalized that our only option is to adapt to their presence. Potential for restoration and rehabilitation depends on the invasive species and system invaded. For example, work on earthworm effects should help to inform management for native plant species of concern in the presence of certain species of nonnative earthworms (e.g., Gundale 2002). In some cases, significant alterations in forest management strategies may be necessary to adapt to the changes caused by earthworm invasion. Forest Service R&D can contribute significantly to such efforts, some of which are already under way in the NRS (Lilleskov, Swanston, Kolka). Certain invasive plants are also known to produce allelopathic chemicals, which can reside in soils long after the aboveground portions of the plant have been treated or removed (Kulmatiski and Beard 2006). The soil-mediated legacy effects of invasive plants are in much need of further research.

Application and Communication

From the current effort it is clear that the Forest Service has a diversity of research programs on a broad range of invasive soil organisms and other species that affect belowground processes. The dispersed structure of our organization constrains communication among the scientists working on these diverse programs and does not emphasize to our customers the diversity and magnitude of our efforts. We suggest that, to improve internal information sharing and to emphasize our strengths in this area to our customers, the Forest Service produce a central Web-based clearinghouse of Forest Service research on soil invasive species on the Forest Service invasive species Web site. We also recommend regular national meetings on invasive soil organisms hosted by the Forest Service. To reduce costs and carbon footprints, some of these meetings should be videoconferences. Establishment of a Forest Service listserv on invasive species would also facilitate communication among Forest Service scientists.

Key Issues

In addition to listing the activities and general recommendations in the previous text, we have identified several key issues in need of coordinated effort in the future.

Identification, Distribution, and Effects of Key Invaders

One of the greatest challenges in characterizing invasive species and their distribution is our lack of thorough taxonomic and biogeographic treatments for many of these groups. Until we have this information it will be impossible to make informed decisions about how to prevent and manage invasions. *Therefore, maintaining and expanding sources of taxonomic and biogeographic knowledge, either within the Forest Service or among our collaborators, should be a key priority of Forest Service R&D.*

Prioritized List of Soil Invasive Species

Given the large number of soil invasive species, it becomes imperative that we prioritize our efforts to address the most pressing issues first. This approach requires a synthesis of expert knowledge on the topic. This document serves as a starting point for such a synthesis and prioritization, but we must be aware that, as our understanding of the actual or potential effects of certain groups increases, our priorities will be likely to shift and that more formalized efforts are necessary to ensure adequate prioritization. *To aid in prioritization efforts, we should initiate a formalized Forest Service R&D prioritization effort (e.g., via a national task force of Forest Service R&D and current and potential academic collaborators on soil invasive species).*

Effects of Invasive Species on Interactions Between Plants and Soil Microbial Communities

At present the feedbacks between invasive and native plant species, soil organisms, and soil microbial communities are poorly understood, yet the potential for large-scale shifts in biodiversity and ecosystem processes driven by symbiotic (whether beneficial or detrimental) organisms is large. To the examples of invasive N-fixing plant species and ectomycorrhizal fungi described previously can be added a host of other interactions. These interactions include the escape of invasive organisms from biological control agents such as root pathogens, resulting in increased competitive advantage for the invasive species (e.g., Callaway et al. 2001, Callaway et al. 2004, Klironomos 2002, Mitchell and Power 2003, Wolfe and Klironomos 2005); the introduction of sublethal root pathogens that will affect plant productivity yet remain undetected; alterations of soil microbial communities by invasive plants (e.g., Kourtev et al. 2002, 2003); and changes in soil microbial communities mediated by the interactions of disturbance and invasive plant species and resulting effects of these changes on native plant regeneration (e.g., Hebel et al. 2009). We face enormous challenges in detection of these interactions, requiring utilization of rapidly developing molecular approaches for characterization of microbial (including both bacterial and fungal) communities. *At present, although local Forest Service R&D efforts on these problems exist (e.g., PNWRS, J.E. Smith), no unified effort is in place for determining a Forest Service strategy to respond to this suite of interrelated problems. Coordination of research efforts among spatially diffuse groups (e.g., via creation of a Forest Service listserv on invasive species) would increase the probability of research coordination among regions.*

Invasive-Dominated Communities in Urban and Agricultural Ecosystems

Areas with higher human population density are exposed to higher loading of introduced soil organisms (Lilleskov et al. 2008). As a result, urban areas have fundamentally different soil arthropod communities with a much higher proportion of introduced species when compared with those in areas of lower human population density (e.g., Bolger et al. 2000, Connor et al. 2002, McIntyre 2000, Pouyat et al. 1994, Spence and Spence 1988). This pattern is likely a function of rates of propagule input, greater disturbance, and other mechanisms such as higher success rates of human-associated faunas because of the match of source and destination habitats. *With the expansion of the wildland-urban interface, the area with dramatically altered communities of soil organisms is also likely to expand. The Forest Service should be prepared to predict the effects of these changes and develop proactive approaches to this problem. The*

NRS (Pouyat) has a program specifically addressing the effects of invasive soil organisms in the urban matrix and exploring soil community and process changes in forests along urban-rural gradients in New York City and Baltimore (Pouyat and Carriero 2003, Steinberg et al. 1997, Szlavecz et al. 2006). A related problem is the interface between wildland and agricultural systems, which could be a major steppingstone for exotic species introductions into wildlands.

One uncertainty is whether urban/human dominated forest soil communities are constrained to their urban matrix or if they will spread into nonurban environments in the future. Patterns of apparent synanthropy (close association with humans) can be driven by slow rates of dispersal or by absolute environmental constraints that limit dispersal into natural ecosystems; i.e., some species may be obligate synanthropes constrained to human-altered environments, whereas others may be facultative synanthropes capable of spreading into ecosystems not dominated by humans (Bolger et al. 2000, Kavanaugh and Erwin 1985, Niemelä and Spence 1991, Niemelä et al. 2002, Spence and Spence 1988). The importance of spread out of these systems will depend on the biotic and abiotic resistance of wildlands to the invaders from urban and agroecosystems (e.g., Hendrix et al. 2006). *To predict future patterns of invasion and community change, it is essential that we determine which key invaders are capable of crossing from human-dominated ecosystems into wildland ecosystems.*

Additive and Synergistic Effects of New Suites of Soil Invasive Species

The concept of an invasional meltdown (i.e., synergistic interactions between invasive species facilitating more invasions and leading to rapid ecosystem change [Simberloff and Von Holle 1999]) has received considerable attention (Simberloff 2006). The potential for invasional meltdown in response to soil invasive species has not been fully explored but has been demonstrated to some degree in interactions between introduced earthworms and introduced plants (Heneghan et al. 2007, Kourtev et al. 1999) and between invasive plants and their mycorrhizal partners (Richardson et al. 2000). Wholesale soil community changes in urban ecosystems and the likely expansion of these invasive-dominated communities highlight the necessity of determining whether such synergistic interactions among invasive species occur among the species in these communities. The Forest Service has researchers with expertise

and active research programs on urban invasive soil organisms and their effects on soil processes (NRS, R. Pouyat), and the expertise of other scientists with relevant interests and skills could be brought to bear on this problem as well. *Better communication, coordination, and collaboration between Forest Service scientists working on invasive plants and those with expertise in soil ecology will help to determine the importance of interactions among invasive species in forest ecosystems.*

Interactions of Invasive Species and Global Environmental Change

Managing in the face of global change is one of the key challenges facing the Forest Service. As climate and atmospheric chemistry change, ecosystems will change in their susceptibility and response to invasions (Dale et al. 2000). Species with direct and indirect effects on soils are likely constrained by the bioclimatic match between source and receptor regions (Lilleskov et al. 2008), so a thorough understanding of how ranges are likely to change as a function of climate change is essential. Other factors, such as nitrogen deposition, provide further stresses on native ecosystems that have a potential for accelerating invasions or increasing their effects (e.g., Fenn et al. 2003). Increasing carbon dioxide will drive alterations in plant tissue chemistry or competitive interactions in ways that could favor or inhibit invasive species. Our understanding of such interactions is poor, but Forest Service research on the interaction of climate and pine beetle outbreaks in the Rocky Mountains (Rocky Mountain Research Station, Logan) indicates the strong nonlinear changes in pest effects that can occur in response to small changes in climate (Logan and Powell 2001). Similar changes could be expected in some invasive soil organisms. For example, invasive native (*Reticulitermes flavipes*) and nonnative (e.g., *Coptotermes formosanus*) termite species are spreading northward but, at present, are apparently limited bioclimatically to the Southern United States and in the North to urban areas (e.g., Arango and Green 2007). Populations resident in urban areas could rapidly expand and coalesce as climate changes. *Understanding the constraints on their distribution and potential for control is critical to our ability to manage and adapt to these invasions.*

Forests are an important sink for carbon, potentially providing an important negative feedback to global climate change. This carbon is stored in both soils and biomass. Changes in soil processes can change the pools of soil carbon. One area in need of investigation is the effect of invasive soil organisms in general, and earthworms in particular, on the storage of carbon in soils. It has been established that short-term losses of soil carbon occur during earthworm invasion. These losses can be significant, on the order of 0.6 Mg C ha^{-1} yr^{-1} (Alban and Berry 1994). How long these losses persist and how these losses are modulated by soil texture is only poorly understood, yet such knowledge is essential to our ability to manage forests for carbon storage. It is possible that, although short-term effects include carbon losses, long-term effects include increased physical and chemical protection of soil organic matter, leading to net accumulations of soil carbon. A Forest Service researcher (NRS, Pouyat) and collaborators have compared the effects of invasive earthworms on soil carbon dynamics in forest stands along an urban-rural gradient in the New York City metropolitan area (Groffman et al. 1995, Pouyat et al. 2002), and an existing Forest Service project addresses this question in nonurban forests (NRS-Lilleskov, Swanston, Kolka). *Understanding the effects of invasive species on climate is critical to our ability to manage forests to mitigate climate change.*

Overarching and Concluding Remarks

The Forest Service has the unique opportunity to provide information that will improve our ability to predict, prevent, detect, eradicate, and manage invasive organisms that affect the belowground component of ecosystems, but only if we make it a priority to develop an integrated program that takes full advantage of the unique strengths that we offer.

These strengths include (1) a national network of capable and motivated scientists dedicated to protecting the sustainability, biodiversity, health, and ecosystem goods and services of our forests; (2) a network of long-term ecological research projects, sites, and associated databases, especially Forest Inventory Analysis and our invaluable network of Long Term Ecological Research Network studies and experimental forests and ranges; (3) a long-term perspective; and (4) the flexibility to respond to problems as they arise. *Our customers—the public, land managers, other scientists, policymakers, other Government organizations such as APHIS—do not have the expertise or focus on forests that would permit them to make informed decisions necessary to protect forests from soil*

invasive species. It is, therefore, our responsibility to ensure that information needed is developed and disseminated to these customers. APHIS needs our help in identifying organisms of greatest concern to forests and their mode of entry, land managers need to recognize the signs of invasive species and know how to respond, other scientists need unfettered access to the information we collect, and the public and policymakers need to be educated about the issues we face and the collective and individual actions required.

To provide this information, we must maintain a vital and comprehensive research program that includes both internal Forest Service research and collaboration with academic institutions and scientific societies. Given the range of taxonomic groups and ecological roles encompassed by soil invasive species, **it is also essential that Forest Service R&D maintain diverse expertise in soil science, pathology, entomology, mycology, taxonomy (especially of invertebrates and fungi), biological and other methods of control, ecology, ecosystem science, biogeochemistry, and hydrology.** The trend toward declining research budgets must be reversed if we are to maintain the expertise needed to address these problems. To address these problems, we need to communicate the economic, social, and environmental benefits of expanding our research program.

We can scale the organization to focus on the highest priority issues/needs if we remain flexible; keep lines of communication open among different research groups; minimize top-down control of the research process, allowing individual scientists to recognize and respond to emerging issues; provide resources for research collaborations among units and regions, and between units and universities; and provide central databases of scientists, expertise, publications, and long-term datasets. It is essential that within Forest Service R&D we share data and protocols to maximize comparability of collected data, and when appropriate seek additional resources to fund collaborative efforts that synthesize existing data or permit national-scale efforts. In order to provide greater integration, it is also critical that we increase incentives, such as targeted funding, for collaboration among regions. Providing opportunities for unstructured communication among researchers (e.g., national meeting of researchers focusing on belowground invasive species, perhaps in partnership with the Soil Ecology Society) will also facilitate such efforts.

Acknowledgments

The authors thank William Mattson, Paul Hendrix, and James Boyle and an anonymous reviewer for feedback on this paper.

Literature Cited

Alban, D.H.; Berry, E.C. 1994. Effects of earthworm invasion on morphology, carbon, and nitrogen of a forest soil. Applied Soil Ecology. 1: 243–249.

Arango, R.; F. Green III. 2007. Feasibility study of eradicating *Reticulitermes flavipes* from Endeavor, Wisconsin. In: Barnes, H.M., ed. Proceedings, Wood Protection 2006. New Orleans, LA: Forest Products Society: 363–368.

Blossey, B.; Hunt-Joshi, T.R. 2003. Belowground herbivory by insects: influence on plants and aboveground herbivores. Annual Review of Entomology. 48: 521–547.

Boag, B.; Yeates, G.W. 2001. The potential impact of the New Zealand flatworm, a predator of earthworms, in Western Europe. Ecological Applications. 11: 1276–1286.

Bohlen, P.J.; Groffman, P.M.; Fahey, T.J., et al. 2004a. Ecosystem consequences of exotic earthworm invasion of north temperate forests. Ecosystems. 7: 1–12.

Bohlen, P.J.; Scheu, S.; Hale, C.M., et al. 2004b. Non-native invasive earthworms as agents of change in northern temperate forests. Frontiers in Ecology and the Environment. 2: 427–435.

Bolger, D.T.; Suarez, A.V.; Crooks, K.R., et al. 2000. Arthropods in urban habitat fragments in Southern California: area, age and edge effects. Ecological Applications. 10: 1230–1248.

Brooks, M.L.; D'Antonio, C.M.; Richardson, D.M., et al. 2004. Effects of invasive alien plants on fire regimes. BioScience. 54: 677–688.

Callaham, M.A., Jr.; Blair, J.M. 1999. Influence of differing land management on the invasion of North American tallgrass prairie soils by European earthworms. Pedobiologia. 43: 507–512.

Callaham, M.A., Jr.; González, G.; Hale, C., et al. 2006a. Policy and management responses to earthworm invasions. Biological Invasions. 8: 1317–1329.

Callaham, M.A., Jr.; Hendrix, P.F.; Phillips, R.J. 2003. Occurrence of an exotic earthworm (*Amynthas agrestis*) in undisturbed soils of the southern Appalachian Mountains, USA. Pedobiologia. 47: 466–470.

Callaham, M.A., Jr.; Richter, D.D.; Coleman, D.C.; Hofmockel, M. 2006b. Long-term land use effects on soil invertebrate communities in Southern Piedmont soils. European Journal of Soil Biology. 42: S150–S156.

Callaway, R.; Newingham, B.; Zabinski, C.A.; Mahall, B.E. 2001. Compensatory growth and competitive ability of an invasive weed are enhanced by soil fungi and native neighbours. Ecology Letters. 4: 429–433.

Callaway, R.M.; Thelen, G.C.; Rodriguez, A.; Hoben, W.E. 2004. Soil biota and exotic plant invasion. Nature. 427: 731–733.

Castellano, M. 2008. Personal communication. Research Forester, Forest Service, Pacific Northwest Research Station, 3200 SW Jefferson Way, Corvallis, OR 97331.

Chen, J.S.-C. 2004. Pathophysiology and transmission of *Thelohania solenopsae* in the red imported fire ant, *Solenopsis invicta*. College Station, TX: Texas A&M University. 138 p. Ph.D. dissertation.

Christian, C.E. 2001. Consequences of a biological invasion reveal the importance of mutualism for plant communities. Nature. 413: 635–639.

Clausen, C.A.; Yang, V.W.; West, M. 2007. Multicomponent biocide composition for wood protection; filed, U.S. Patent Office, March 20, 2007.

Connor, E.F.; Hafernik, J.; Levy, J., et al. 2002. Insect conservation in an urban biodiversity hotspot: the San Francisco Bay area. Journal of Insect Conservation. 6: 247–259.

Dale, V.H.; Joyce, L.A.; McNulty, S.; Neilson, R.P. 2000. The interplay between climate change, forests, and disturbances. The Science of the Total Environment. 262: 201–204.

Dangerfield, J.M.; McCarthy, T.S.; Ellery, W.N. 1998. The mound-building termite *Macrotermes michaelseni* as an ecosystem engineer. Journal of Tropical Ecology. 14: 507–520.

Ducey, P.K.; Messere, M.; Lapoint, K.; Noce, S. 1999. Lumbricid prey and potential herpetofaunal predators of the invading terrestrial flatworm *Bipalium adventitium* (Turbellaria: Tricladida: Terricola). The American Midland Naturalist. 141: 305–314.

Ducey, P.K.; Noce, S. 1998. Successful invasion of New York State by the terrestrial flatworm, *Bipalium adventitium*. Northeastern Naturalist. 5: 199–206.

Ehrenfeld, J.G. 2003. Effects of exotic plant invasions on soil nutrient cycling processes. Ecosystems. 6: 503–523.

Ellison, A.M.; Bank, M.S.; Clinton, B.D., et al. 2005. Loss of foundation species: consequences for the structure and dynamics of forested ecosystems. Frontiers in Ecology and the Environment. 3: 479–486.

Feener, D.H., Jr. 2000. Is the assembly of ant communities mediated by parasitoids? Oikos. 90: 79–88.

Fenn, M.E.; Baron, J.S.; Allen, E.B., et al. 2003. Ecological effects of nitrogen deposition in the Western United States. BioScience. 53: 404–420.

Fiore, C.; Tull, J.L.; Zehner, S.; Ducey, P.K. 2004. Tracking and predation on earthworms by the invasive terrestrial planarian *Bipalium adventitium* (Tricladida, Platyhelminthes). Behavioural Processes. 67: 327–334.

Frelich, L.E.; Hale, C.M.; Scheu, S., et al. 2006. Earthworm invasions into previously earthworm-free temperate and boreal forests. Biological Invasions. 8: 1235–1245.

Gange, A.C.; Brown, V.K. 2002. Soil food web components affect plant community structure during early succession. Ecological Research. 17: 217–227.

González, G. 2002. Soil organisms and litter decomposition. In: Ambasht, R.S.; Ambasht, N.K., eds. Modern trends in applied terrestrial ecology. New York: Kluwer Academic/Plenum Publishers: 315–329.

González, G. 2006. Earthworms as invasive species in Latin America—the 2nd Latin American Meeting on *Oligochaeta* (Earthworm) Ecology and Taxonomy (Editorial). Caribbean Journal of Science. 42(3): 281–284.

González, G.; Espinosa, E.; Zhigang, L.; Zou, X. 2006a. A fluorescent marking and re-count technique using the invasive earthworm, *Pontoscolex corethrurus* (Annelida: Oligochaeta). Caribbean Journal of Science. 42(3): 371–379.

González, G.; García, E.; Cruz, V., et al. 2007a. Earthworm communities along an elevation gradient in Northeastern Puerto Rico. European Journal of Soil Biology. 43: 24–32.

González, G.; Huang, C.-Y.; Chuang, S.C. 2008. Earthworms and post-agricultural succession. In: Myster, R.W., ed. Post-agricultural succession in the Neotropics. New York: Springer: 115–138.

González, G.; Huang, C.-Y.; Zou, X.; Rodríguez, C. 2006b. Earthworm invasions in the tropics. Biological Invasions. 8: 1247–1256.

González, G.; Seastedt, T.R. 2001. Soil fauna and plant litter decomposition in tropical and subalpine forests. Ecology. 82: 955–964.

González, G.; Seastedt, T.R.; Donato, Z. 2003. Earthworms, arthropods and plant litter decomposition in aspen (*Populus tremuloides*) and lodgepole pine (*Pinus contorta*) forests in Colorado, USA. Pedobiologia. 47: 863–869.

González, G.; Zou, X. 1999. Plant and litter influences in earthworm abundance and community structure in a tropical wet forest. Biotropica. 31: 486–493.

González, G.; Zou, X.; Borges, S. 1996. Earthworm abundance and species composition in abandoned tropical croplands: comparison of tree plantations and secondary forests. Pedobiologia. 40: 385–391.

González, G.; Zou, X.; Li, Y. 2007b. Effects of post-hurricane fertilization and debris removal on earthworm abundance and biomass in subtropical forests in Puerto Rico. In: Brown, G.G.; Fragoso, C., eds. Minhocas na América Latina: biodiversidade e ecologia. Londrina, Brazil: EMBRAPA Soja: 99–108.

González, G.; Zou, X.; Sabat, A.; Fetcher, N. 1999. Earthworm abundance and distribution pattern in contrasting plant communities within a tropical wet forest in Puerto Rico. Caribbean Journal of Science. 35: 93–100.

Greenup, M. 1998. Managing *Chamaecyparis lawsoniana* (Port-Orford-Cedar) to control the root disease caused by *Phytophthora lateralis* in the Pacific Northwest, USA. In: Laderman, X., ed. Coastally restricted forests. Oxford, United Kingdom: Oxford University Press: 93–101.

Groden, E.; Drummond, F.A.; Garnas, J.; Franceour, A. 2005. Distribution of an invasive ant, *Myrmica rubra* (Hymenoptera: Formicidae), in Maine. Journal of Economic Entomology. 98: 1774–1784.

Groffman, P.M.; Pouyat, R.V.; McDonnell, M.J., et al. 1995. Carbon pools and trace gas fluxes in urban forest soils. In: Lal, R.; Kimble, J.; Levine, E.; Stewart, B.A., eds. Soil management and greenhouse effect. Boca Raton, FL: CRC Press: 147–157.

Gundale, M.J. 2002. Influence of exotic earthworms on the soil organic horizon and the rare fern *Botrychium mormo*. Conservation Biology. 16: 1555–1561.

Hale, C.; Frelich, L.; Reich, P.; Pastor, J. 2005. Effects of European earthworm invasion on soil characteristics in northern hardwood forests of Minnesota, USA. Ecosystems. 8: 911–927.

Hale, C.M.; Frelich, L.E.; and Reich, P.B. 2006. Changes in hardwood forest understory plant communities in response to European earthworm invasions. Ecology. 87: 1637–1649.

Hebel, C.L.; Smith, J.E.; Cromack, K., Jr. 2009. Invasive plant species and soil microbial response to wildfire burn severity in the Cascade Range of Oregon. Applied Soil Ecology. 42: 150–159.

Hendrix, P.F.; Baker, G.; Callaham, M.A. Jr., et al. 2006. Invasion of exotic earthworms into ecosystems inhabited by native earthworm communities. Biological Invasions. 8: 1287–1300.

Heneghan, L.; Steffen, J.; Fagen, K. 2007. Interactions of an introduced shrub and introduced earthworms in an Illinois urban woodland: impact on leaf litter decomposition. Pedobiologia. 50: 543–551.

Hickerson, C.-A.M.; Anthony, C.D.; Walton, B.M. 2005. Edge effects and intraguild predation in native and introduced centipedes: evidence from the field and from laboratory microcosms. Oecologia. 146: 110–119.

Holdsworth, A.R.; Frelich, L.E.; Reich, P.B. 2007. Effects of earthworm invasion on plant species richness in northern hardwood forests. Conservation Biology. 21: 997–1008.

Holway, D.A.; Lach, L.; Suarez, A.V., et al. 2002. The causes and consequences of ant invasions. Annual Review of Ecology and Systematics. 33: 181–233.

Huang, C.Y.; González, G.; Hendrix, P.F. 2006. The recolonization ability of a native earthworm species, *Estherella* spp, in Puerto Rican forests and pastures. Caribbean Journal of Science. 42(3): 386–396.

Hunter, M.D. 2001. Out of sight, out of mind: the impacts of root-feeding insects in natural and managed systems. Agricultural and Forest Entomology. 3: 3–9.

Jass, J.; Klausmeier, B. 2000. Endemics and immigrants: North American terrestrial isopods (Isopoda, Oniscidea) north of Mexico. Crustaceana. 73: 771–799.

Jones, C.G.; Lawton, J.H.; Shachak, M. 1994. Organisms as ecosystem engineers. Oikos. 69: 373–386.

Jurgensen, M.F.; Storer, A.J.; Risch, A.C. 2005. Red wood ants in North America. Annales Zoologici Fennici. 42: 235–242.

Kavanaugh, D.H.; Erwin, T.L. 1985. *Trechus obtusus* Erichson (Coleoptera: Carabidae), a European ground beetle, on the Pacific Coast of North America: its distribution, introduction, and spread. Pan-Pacific Entomologist. 61: 170–179.

Klironomos, J. 2002. Feedback with soil biota contributes to plant rarity and invasiveness in communities. Nature. 417: 67–70.

Kotliar, N.B. 2000. Application of the new keystone-species concept to prairie dogs: How well does it work? Conservation Biology. 14: 1715–1721.

Kourtev, P.S.; Ehrenfeld, J.G.; Haggblom, M. 2002. Exotic plant species alter the microbial community structure and function in the soil. Ecology. 83: 3152–3166.

Kourtev, P.S.; Ehrenfeld, J.G.; Haggblom, M. 2003. Experimental analysis of the effect of exotic and native plant species on the structure and function of soil microbial communities. Soil Biology and Biochemistry. 35: 895–905.

Kourtev, P.S.; Huang, W.Z.; Ehrenfeld, J.G. 1999. Differences in earthworm densities and nitrogen dynamics in soils under exotic and native plant species. Biological Invasions. 1: 237–245.

Kulmatiski, A.; Beard, K.H. 2006. Activated carbon as a restoration tool: potential for control of invasive plants in abandoned agricultural fields. Restoration Ecology. 14: 251–257.

Larson, D.J.; Langor, D.W. 1982. The carabid beetles of insular Newfoundland (Coleoptera: Carabidae: Cicindellidae)—30 years after Lindroth. Canadian Entomologist. 114: 594–597.

Lawrence, B.; M. C. Fisk; T. J. Fahey; Suarez, E.R. 2003. Influence of nonnative earthworms on mycorrhizal colonization of sugar maple (*Acer saccharum*). New Phytologist. 157: 145–153.

Lax, A.R.; Osbrink, W.L. 2003. United States Department of Agriculture—Agriculture Research Service research on targeted management of the Formosan subterranean termite *Coptotermes formosanus* Shiraki (Isoptera: Rhinotermitidae). Pest Management Science. 59: 788–800.

Lebow, S.; Shupe, T.; Woodward, B., et al. 2006. Formosan and native subterranean termite attack of pressure-treated SPF wood species exposed in Louisiana. Wood and Fiber Science. 38: 609–620.

Leung, B.; Lodge, D.M.; Finnoff, D., et al. 2002. An ounce of prevention or a pound of cure: bioeconomic risk analysis of invasive species. Proceedings of the Royal Society B: Biological Sciences. 269: 2407–2413.

Lilleskov, E.A.; Mattson, W.J.; Storer, A.J. 2008. Divergent biogeography of native and introduced soil macroinvertebrates in North America north of Mexico. Diversity and Distributions. 14: 893–904.

Lilleskov, E.A.; Parrent, J.L. 2007. Can we develop general predictive models of mycorrhizal fungal community-environment relationships? New Phytologist. 174: 250–256.

Lodge, D.J. 2001. Diversidad mundial y regional de hongos. In: Hernández, H.M.; García-Aldrete, A.; Alvarez, F.; Ulloa, M., eds. Enfoques Contemporáneos para el Estudio de la Biodiversity. Ediciones Científicas Universitarias, Serie Texto Científico Universitario, Instituto de Biología, UNAM, Ciudad Universitario, Mexico: 291–304.

Logan, J.; Powell, J. 2001. Ghost forests, global warming, and the mountain pine beetle (Coleoptera: Scolytidae). American Entomologist. 47: 160–172.

Masters, G.J. 2004. Belowground herbivores and ecosystem processes. In: Weisser, W.W.; Siemann, E., eds. Ecological studies, vol. 173. Insects and ecosystem function. Berlin, Germany: Springer: 193–204.

McIntyre, N.E. 2000. Ecology of urban arthropods: a review and a call to action. Annals of the Entomological Society of America. 93: 825–835.

Mitchell, C.E.; Power, A.G. 2003. Release of invasive plants from fungal and viral pathogens. Nature. 421: 625–627.

Mueller, G.M.; Schmit, J.P.; Leacock, P.R., et al. 2007. Global diversity and distribution of macrofungi. Biodiversity and Conservation. 16: 37–48.

Niemelä, J.; Kotze, D.J.; Venn, S., et al. 2002. Carabid beetle assemblages (Coleoptera, Carabidae) across urban-rural gradients: an international comparison. Landscape Ecology. 17: 387–401.

Niemelä, J.; Spence, J.R. 1991. Distribution and abundance of an exotic ground-beetle (Carabidae): a test of community impacts. Oikos. 62: 351–359.

Ogren, R.E.; Kawakatsu, M. 1998. American Nearctic and Neotropical land planarian (Tricladida: Terricola) faunas. Pedobiologia. 42: 441–451.

Ortiz-Santana, B. 2006. Phylogeny and biogeography of Caribbean Boletales. Rio Piedras, PR: University of Puerto Rico at Rio Piedras, Faculty of Natural Sciences. 305 p. Ph.D. dissertation.

Peterson, A.T.; Vieglais, D.A. 2001. Predicting species invasions using ecological niche modeling: new approaches from bioinformatics attack a pressing problem. Bioscience. 51: 363–371.

Pinski, R.A.; Mattson, W.J.; Raffa, K.F. 2005a. Composition and seasonal phenology of a nonindigenous root-feeding weevil (Coleoptera: Curculionidae) complex in northern hardwood forests in the Great Lakes Region. Environmental Entomology. 34: 298–307.

Pinski, R.A.; Mattson, W.J.; Raffa, K.F. 2005b. Host breadth and ovipositional behavior of adult *Polydrusus sericeus* and *Phyllobius oblongus* (Coleoptera: Curculionidae), nonindigenous inhabitants of northern hardwood forests. Environmental Entomology. 34: 148–157.

Pouyat, R.; Groffman, P.; Yesilonis, I.; Hernandez, L. 2002. Soil carbon pools and fluxes in urban ecosystems. Environmental Pollution. 116: S107–S118.

Pouyat, R.V.; Carreiro, M.M. 2003. Contrasting controls on decomposition of oak leaf litter along an urban-rural land use gradient. Oecologia. 135: 288–298.

Pouyat, R.V.; Parmelee, R.W.; Carreiro, M.M. 1994. Environmental effects of forest soil-invertebrate and fungal densities in oak stands along an urban-rural land use gradient. Pedobiologia. 38: 385–399.

Reichard, S.H.; Hamilton, C.W. 1997. Predicting invasions of woody plants introduced into North America. Conservation Biology. 11: 193–203.

Reynolds, J.W.; Wetzel, M.J. 2004. Terrestrial Oligochaeta (Annelida: Clitellata) in North America north of Mexico. Megadrilogica. 9: 71–98.

Richardson, D.M.; Allsopp, N.; D'Antonio, C.M., et al. 2000. Plant invasions—the role of mutualisms. Biological Reviews. 75: 65–93.

Rojas, M.; Morales-Ramos, J.; Green, F., III. 2004. Naphthalenic compounds as termite bait toxicants. U.S. Patent # 6,691,453.

Scheu, S. 2001. Plants and generalist predators as links between the below-ground and above-ground system. Basic and Applied Ecology. 2: 3–13.

Schwartz, M.W.; Hoeksema, J.D.; Gehring, C.A., et al. 2006. The promise and the potential consequences of the global transport of mycorrhizal fungal inoculum. Ecology Letters. 9: 501–515.

Simberloff, D. 2006. Invasional meltdown 6 years later: important phenomenon, unfortunate metaphor, or both? Ecology Letters. 9: 912–919.

Simberloff, D.; Stiling, P. 1996. How risky is biological control? Ecology. 77: 1965–1974.

Simberloff, D.; Von Holle, B. 1999. Positive interactions of nonindigenous species: Invasional meltdown? Biological Invasions. 1: 21–32.

Singer, R. 1981. Wild pig populations in the National Parks. Environmental Management. 5: 263–270.

Spence, J.R.; Spence, D.H. 1988. Of ground-beetles and men: introduced species and the synanthropic fauna of western Canada. Memoirs of the Entomological Society of Canada. 144: 151–168.

Steinberg, D.A.; Pouyat, R.V.; Parmelee, R.W.; Groffman, P.M. 1997. Earthworm abundance and nitrogen mineralization rates along an urban-rural land use gradient. Soil Biology and Biochemistry. 29: 427–430.

Stevens, G.N.; Jones, R.H.; Mitchell, R.J. 2002. Rapid fine root disappearance in a pine woodland: a substantial carbon flux. Canadian Journal of Forest Research. 32: 2225–2230.

Stohlgren, T.J.; Barnett, R.H.; Kartesz, J.T. 2004. The rich get richer: patterns of plant invasions in the United States. Frontiers in Ecology and the Environment. 1: 11–14.

Sugimoto, A.; Inoue, T.; Tayasu, I., et al. 1998. Methane and hydrogen production in a termite-symbiont system. Ecological Research. 13: 241–257.

Szlavecz, K.; Placella, S.A.; Pouyat, R.V., et al. 2006. Invasive earthworm species and nitrogen cycling in remnant forest patches. Applied Soil Ecology. 32: 54–62.

Tupen, J.; Roth, B. 2001. Further spread of the introduced decollate snail, *Rumina decollata* (Gastropoda: Pulmonata: Subulinidae), in California, USA. The Veliger. 44: 400–404.

Van der Putten, W.H.; Vet, L.E.M.; Harvey, J.A.; Wackers, F.L. 2001. Linking above- and belowground multitrophic interactions of plants, herbivores, pathogens, and their antagonists. Trends in Ecology & Evolution. 16: 547–554.

Vitousek, P.M.; Walker, L.R. 1989. Biological invasion by *Myrica faya* in Hawaii: plant demography, nitrogen fixation, and ecosystem effects. Ecological Monographs. 59: 247–265.

Wells, C.E.; Glenn, D.M.; Eissenstat, D.M. 2002. Soil insects alter fine root demography in peach (*Prunus persica*). Plant, Cell & Environment. 25: 431–439.

Wheeler, G.S.; Tokoro, M.; Scheffrahn, R.H.; Su, N.-Y. 1996. Comparative respiration and methane production rates in Nearctic termites. Journal of Insect Physiology. 42: 799–806.

Wolfe, B.E.; Klironomos, J.N. 2005. Breaking new ground: soil communities and exotic plant invasion. Bioscience. 55: 477–488.

Yiqing, L.; González, G. 2008. Soil fungi and macrofauna in the Neotropics. In: Myster, R.W., ed. Post-agricultural succession in the Neotropics. New York: Springer: 93–114.

Zaborski, E.R. 2002. Observations on feeding behavior by the terrestrial flatworm *Bipalium adventitium* (Platyhelminthes: Tricladida: Terricola) from Illinois. The American Midland Naturalist. 148: 401–408.

Zobel, D.B.; Roth, L.F.; Hawk, G.M. 1982. Ecology, pathology, and management of Port-Orford-cedar (*Chamaecyparis lawsoniana*). Gen. Tech. Rpt. PNW-184. Portland, OR: U.S. Department of Agriculture, Forest Service, Pacific Northwest Forest and Range Experimental Station. 161 p.

Zou, X.; González, G. 1997. Changes in earthworm density and community structure during secondary succession in abandoned tropical pastures. Soil Biology and Biochemistry. 29: 627–629.

Prevention

Kerry Britton[1], Barbara Illman[2], and Gary Man[3]

Abstract

Prevention is considered the most cost-effective element of the Forest Service Invasive Species Strategy (USDA Forest Service 2004). What makes prevention difficult is the desire to maximize free trade and the resulting benefits to society while, at the same time, protecting natural resources. The role of science is to first identify which commodities pose an unacceptable risk and then to either develop mitigations that will adequately ameliorate the risk or technically justify and facilitate commodity exclusion.

Action Items

Developing a system of data sharing that allows a rapid synthesis of information from global databases is a top priority. Keeping pests out requires knowing what is out there, which pathways are contaminated, and how to mitigate the risk. It is costly and time-consuming to develop new science upon the detection of an introduced pest and to eradicate it. Frequently an introduced pest becomes established before management tools are developed, and eradication is no longer possible. To help focus these efforts, global partnerships are needed to identify groups of pests that are of most concern, in advance of their arrival and establishment.

Decision-support tools are needed to help managers decide which commodities pose an unacceptable risk. The decision relies on estimates of potential pest impact and risk mitigation methods effective against pests that are not currently in the United States or that are present in limited distribution and under regulatory control.

Economic impact assessment modules for ecological services valuation need to be developed to inform these decision-support tools. Gaming scenario approaches would help regulators and managers decide which commodities to exclude and which established pests to fight. Cost/benefit analyses are needed to compare mitigation options.

Systematics expertise and pest biology research necessarily underlie the development of these tools. In many cases, information about related species must be used as a proxy for species-specific information on biology, behavior, and risk. Similar species' biologies often provide a starting point for assessing invasiveness or for developing detection tools, commodity mitigation recommendations, and eradication or control strategies. A greater understanding of the response relationships among organisms within taxonomic or biofunctional groups is required to trust the strength of such proxy options. Recent advances have improved our understanding of systematic relationships but also indicate that regulatory concepts based on morphological pest species concepts, in some cases, have afforded inadequate protection from new and more virulent races or subspecies. Quantification of intraspecific variability in key biological attributes such as aggressiveness could help regulators decide what level of taxonomic specificity is appropriate to regulate.

Detection methods are needed to verify pest presence or mitigation success. These methods are needed offshore in export programs, including Animal and Plant Health Inspection Service (APHIS) preclearance programs for high-risk commodities, and in domestic programs. Participation and full support of regional (e.g., North American Plant Protection Organization) and international (e.g., International Plant Protection Convention [IPPC]) initiatives to develop globally accepted detection tools are recommended.

Monitoring tools are needed that include better trapping and baiting methods and scientific assessments of how to most effectively and efficiently deploy those methods to maximize the likelihood of early detection for rapid response.

[1] National Forest Pathology Program Leader, Forest Service, Research and Development, Washington Office, 1601 N. Kent Street, RPC-4, Arlington, VA 22209.

[2] Director, Institute for Microbial and Biochemical Technology, Forest Service, Forest Products Laboratory.

[3] Forest Health Specialist, Forest Service, Forest Health Protection, Forest Health Specialist, Forest Health Protection, 1601 N. Kent Street, RPC-7, Arlington, VA 22209.

Mitigation and control strategies must be developed and improved as new technologies become available. In support of both import and export trade, mitigation strategies reduce the risk of the movement of pests on a particular commodity to an acceptable level. Control strategies reduce or eliminate an established pest population.

Background

Prevention stands as the first, most effective step of the Forest Service National Invasive Species Strategy (USDA Forest Service 2004). Prevention has long been recognized as an important component of bilateral and multilateral trade agreements and international law, but, as the ecological consequences of science gaps have become more obvious, the need to address these gaps has become more widely recognized. The expert panel that conducted the Forest Service Invasive Species Research External Peer Review in October 2006, recommended that Forest Service Research and Development (R&D) put more emphasis on tools to aid in prevention.

Several major pathways have been identified as sources for the introduction of invasive species. Ballast water release from ships is a source of exotic snakes, snails, fish, and mussels. Wood packing material (WPM), such as pallets, boxes, and crates, is a pathway for invasive species such as the Asian long-horned beetle and wood-inhabiting fungi. Wood-boring beetles have also been detected in artificial plants, furniture, and other nontreated wood products. Plants for planting can be pests in and of themselves, or they may provide a pathway for invasive fungi, insects, and nematodes. Escapes and releases from the exotic pet trade and horticultural smuggling provide pathways for a variety of pests.

Trends in Trade

Imports of agricultural, fish, and forestry products are now more than $100 billion a year. Wood product imports increased 45 percent between 2002 and 2006, and exports increased 27 percent during the same period (USDA Foreign Agricultural Service 2007). Such globalization carries with it the burden of increased opportunities for invasive species introduction.

WPM supports 75 percent of goods in transit. The threat posed by this enormous volume of WPM has been partially mitigated by the adoption of rules requiring heat treatment or fumigation, based on an international standard called ISPM-15. Forest Service research played a large part in documenting the need to mitigate the risk associated with WPM, defining effective treatments, and determining the relevance of residual bark in the potential for infestation of WPM after treatment. Research is still needed to test treatment efficacy against thermophilic insects and microbes, to resolve questions about the effect of moisture content on methyl bromide efficacy, and to develop alternatives to methyl bromide, which is at present the only internationally approved fumigant. APHIS recently issued a supplement to its Environmental Impact Assessment for use of methyl bromide as a treatment for WPM, with revised estimates of between 744 and 2,110 metric tons per year (USDA APHIS 2007). A notice of decisionmaking published in February 2008 advised that, because of the absence of accepted alternatives, this ozone-depleting compound would continue to be accepted in lieu of heat treatment.

Bergman, Chandler, and Locklear (2002) predicted that the number of invasive vertebrate species in the United States will continue to increase due to smuggling, species escapes from confinement, and minimal border inspections. They summarized APHIS Wildlife Services' (WS's) records of cooperator estimates of economic losses, with the caveat that monetary losses reported to WS are miniscule when compared to the overall damages caused by these species. "Much of the damage, such as losses of threatened and endangered species, and damage to natural areas is not readily quantifiable. The true picture of the harm caused by invasive species cannot be fully appreciated until additional research is conducted and better models are developed to determine the full amount of damage caused by individual invasive species and groups" (Bergman et al. 2002).

Imports of live plants increased from 1 billion in 2004 to 2.5 billion in 2006. Most (75 percent) pass through the port of Miami, where 34 inspectors struggle to keep pace with the increasing deluge. Relying on inspection to detect pests is inadequate because not all invasive insects and pathogens that were introduced via nursery stock imports were known to science before their establishment in the United States. Although in the past plants were usually imported for breeding, or as mother plants for domestic production, they are increasingly being produced offshore for retail sales. Thus, the associated pests go straight to the consumer, rather than remaining under the watchful eye of a plant breeder or nursery worker. APHIS is currently revising its regulations for "plants for planting," and this process is expected to take a decade to complete. Because APHIS does not have a research arm, the USDA Agricultural Research Service and Forest Service R&D need to provide the science to help inform this process. The intent of this regulation is to reduce unintentional pest introductions on the plants and to institute a process to screen intentionally introduced plants for invasiveness.

Another trend in trade is the increasing use of containers for shipping. As containers pass through port inspection stations, fewer than 2 percent are inspected. The other 98 percent are not opened until they reach their final destination, which has made every receiving town in America a potential host to hitchhikers, such as the Asian gypsy moth, other pests associated with WPM, and commodity-associated pests.

Science Challenges

Prevention is considered the least expensive and most effective approach to reducing the effects of invasive species; however, the "Agreement on the Application of Sanitary and Phytosanitary Measures" of the World Trade Organization (WTO) (WTO 1994) requires that every regulation be "based on scientific principles and…not maintained without sufficient scientific evidence." The WTO recognizes the IPPC as the body to harmonize standards for plant health. The IPPC stipulates that countries are not allowed to require phytosanitary measures against nonregulated pests. Both conditions imply a level of knowledge about pests that is frequently lacking. In fact, the invasiveness of nonnative forest pests has rarely been predicted in advance. A more systematic approach toward identifying future threats is sorely needed. Forest Service R&D has a significant role to play in helping to develop a holistic, international approach.

Participating in the global discussion surrounding these critical issues and continuing to provide research answers to regulators' questions are essential to redeem Forest Service responsibilities in protecting the Nation's natural resources.

Research Needs

The science of invasion biology is developing rapidly. Although Forest Service R&D has played a pivotal role in this development, with crosscutting studies for individual species,

future program direction should require more organizational emphasis on prevention methods, specifically those described in the following text.

Pathway Assessments

Most scientists and regulators agree that a pathway approach to pest regulation is more effective than a pest-by-pest, country-by-country approach; however, the IPPC has stringent ground rules for pathway pest risk assessments. Risk is assessed for a suite of known pests, and mitigation treatments must be designed to address these specific examples. Although this regulatory approach does not directly address unknown pests (IUFRO 2008) mitigations designed to reduce known pests will reduce the risk of at least some unknown pests as well. Clearly, research is needed to identify the riskiest pathways and the pests likely to use them and to develop mitigation measures that will address them. The success of this pathway approach needs to be quantified.

Exclusion methods based on pathways would concentrate efforts on high-risk, high-volume trade commodities. Risk assessments are needed for global transport of wood chips, peeler cores, and wooden handicraft items. An online database for entry pathway information is needed to list intentional and accidental introduction pathways, provide links to updated pathway risk assessments, identify expected invasive species groups on cargo goods known to provide pathways, and provide links to control and mitigation methods for known pathways. Quite a few individual efforts have begun to build valuable online databases. As Web-based information proliferates, however, it becomes more difficult to find the most credible, condensed information. Forest Service R&D should work with other members of the National Invasive Species Council to unify databases and provide a data-mining portal that links all the relevant sources of information on invasive species.

Many independent sources of data exist regarding offshore pests, but these sources need to be synthesized to help predict potential problem areas. Forest Service scientists should help define the risks from source regions that ship live plants or other commodities likely to harbor invasive species. Climate-matching and forest-type comparisons of regions that present new potential sources of high-risk commodities should be studied in the light of port interceptions and commodity import trends. For example, China and India are said to be "gearing up" to enhance their exports of plants for planting. Collaborative projects should examine pests known to occur in these lands that are likely to occur on proposed commodities, their risk of establishment and potential detrimental effects in the United States.

Mitigation Tools

After potential pests are characterized, it is possible that mitigation measures can be developed. In the case of live plant imports, mitigation measures will be needed to support the implementation of the North American Plant Protection Organization's (NAPPO) Plants for Planting standard adopted in October 2005 (NAPPO 2005) To implement this standard, APHIS will be proposing a systems approach, to which Forest Service R&D should contribute sound science. Specifically, we can help identify "plant genus/source region" combinations that present unacceptable risks to forest and rangeland ecosystems. Such plants would be added to a new category of plants for planting labeled "NAPPRA—not authorized pending pest risk assessment." Later, pest risk analyses would identify mitigation measures that might allow the plant genus to be safely imported from the listed region(s).

Safe trade in wood products requires substantial knowledge of pest biology, and Forest Service scientists must continue their significant role in filling important knowledge gaps. Although invasive insects are well documented in their extensive use of the WPM and wood product pathway, little research on the risks from pathogens associated with wood has been conducted. Despite this dearth of research, a recent study reported that *Phytophthora ramorum* exists in xylem of affected oaks. Debarking is no longer a sufficient treatment; new mitigations are needed to reduce the risk of spreading the pathogen via firewood and other wood products from areas affected by Sudden Oak Death disease. The risks associated with other wilt pathogens and wood decay organisms need to be examined. Domestic movement of firewood has recently been blamed for spreading several important insect pests (e.g., emerald ash borer and Asian longhorned beetle). Research should investigate the risks associated with this pathway and develop acceptable mitigation measures. Furthermore, Canadian wood products cross our border with much less restriction than those of other countries, partly because of a perception that pests will move on their own despite regulatory efforts and partly from the idea that our ecosystems are so similar that our existing pests are much the same. These ideas are researchable hypotheses, rather than proven facts.

Economic Impact Assessments

The economic consequences of invasive species are not well known. Examples for aquatics and estimates for timber pests indicate that the costs are high, ranging in the millions to billions of dollars per year. A systematic assessment of the costs for prevention vs. control of invasive species is needed. The global trade model needs to be expanded to predict some of the economic consequences of invasive species, both social and ecological.

Decisionmakers constantly need information to guide sound choices about which battles to fight. Because of the nature of new pest incursions, regulatory decisions are usually made in the face of great uncertainty. Sensitivity analyses should be conducted to identify key knowledge gaps in our ability to predict pest behavior in this country, based on host preferences and other biological characteristics expressed in the country of origin. How can we predict economic effects in time to inform decisions the regulators must make soon after discovery?

It is essential that Forest Service R&D retain the flexibility to address immediate needs as they arise, providing sound science to support the regulatory community. What exit strategies support the often necessary decision to transition from eradication to slow-the-spread programs?

Containment Measures

Another aspect of prevention is our efforts to prevent the spread of pests already present, but restricted in distribution in the United States. The resources required to maintain effective monitoring and apply control treatments dictate careful targeting of these efforts. Risk assessment of vulnerable ecosystems requires more specific predictive tools for pests under current conditions. The potential effects of climate change on pest biology must also be understood to help prioritize management options. Accurate predictive models rely on studies of pest biology and interactions with host and environment, strengths of Forest Service R&D.

Monitoring Tools

Better detection methods, such as polymerase chain reaction array chips, are needed to help inspectors at ports identify pests associated with risky commodities such as plants for planting. The Barcode of Life project will also help alleviate the dire need for more taxonomists. Forest Service R&D should provide input to this worthy cause to the full extent possible. The barcodes are only as functional as the taxonomic accuracy of the original identifications.

Robot insect sorters are being developed that optically assess diagnostic features, linking into diagnostic information systems. Designed to help ameliorate the loss of taxonomic expertise, this science is in its infancy. Forest Service R&D taxonomic expertise will be required as the mechanics and programming mature and the range of taxa covered by robot sorters expands.

Offshore monitoring programs, such as the very successful Asian gypsy moth program, should be expanded to include many other pests. Trap design and lures for foreign insect pests will be needed, and sentinel plantings of American host species should be established and monitored overseas. This effort will require developing extensive networks of scientists working across borders to identify potential pest problems before they arrive. International databases or data mining systems that link existing national data are sorely needed. These networks will also be extremely beneficial in developing rapid responses.

Despite our best efforts, pests will undoubtedly escape our vigilance and establish new populations in the United States. Tools will always be needed in a short timeframe to provide for detection, faster delineation, and containment of new infestations. In addition to other diagnostic methods, detection via remote sensing will become increasingly important. In the distant future, informatics systems will build cohesive services across agencies, linking inputs from a multitude of databases (e.g., Port Inspection Stations, Cooperative Agricultural Pest Survey, National Plant Diagnostic Network, Early Detection and Rapid Response trapping program, Forest Inventory and Analysis, and Forest Health Monitoring program). Faster detection and better information sharing will help agencies coordinate rapid responses.

Both APHIS and Forest Health Protection (FHP) have proposed to develop citizen-monitoring networks. The USDA agencies should work together to develop the most efficient training and outreach programs and a simplified reporting structure for citizens. Quick, easy-to-use, and inexpensive diagnostic tools are needed to help engage citizens in monitoring. One tool now on the horizon is the species-specific ELISA kit. More accessible, Web-based diagnostic tools are needed. Within the planning timeframe, other methods will also be developed to meet this new need.

For example, Forest Service R&D and FHP are developing and refining software programs and models to help quantify the value of urban forests (including those factors that can devalue an urban forest, such as insects and diseases). Models such as UFORE and STRATUM calculate the amount of air pollutants removed from the atmosphere, carbon stored by urban forests, and the effect of trees on energy use in buildings. More importantly, the models calculate the potential economic effect of an insect or disease attack in urban trees. Other tools such as MCITI collect data that are fed into the models and that can also be used as a diagnostic tool to highlight possible pest problems. Further research and development is needed to make these tools and models more robust and useful to land managers and policymakers.

Technology Transfer and Outreach

The future will see great strides in program delivery via the Internet. Forest Service R&D must not fail to invest in developing user-friendly products that deliver information users need in a format they can use. Strengthening our role in the National Invasive Species Council (NISC) and its plant, aquatic, and terrestrial committees would increase leverage significantly. Educational programs targeting air passengers, pet owners, and horticultural enthusiasts will be more cost-effective if conducted by an interagency action committee under the guidance of NISC. To accomplish our prevention goals, Forest Service R&D clearly needs to expand its influence through better partnerships for implementation and outreach.

Literature Cited

Bergman, D.L.; Chandler, M.D.; Locklear, A. 2002. The economic impact of invasive species to wildlife services' cooperators. In: Clark, L.: Hone J.; Shrivik, J.A., et al., eds. Human conflicts with wildlife: economic considerations. Proceedings of the Third NWRC Special Symposium. Fort Collins, CO: National Wildlife Research Center. 178 p. http://www.aphis.usda.gov/wildlife_damage/nwrc/symposia/economics_symposium/index.shtml.

International Union of Forest Research Organizations (IUFRO). 2008. Recommendation of a pathway approach for regulation of plants for planting: a concept paper from the IUFRO unit on alien invasive species and international trade (adopted by consensus of members). Vienna, Austria: International Union of Forest Research Organizations http://www.forestry-quarantine.org/Documents/IUFRO-ConceptPaper-%20Plants-Planting.pdf.

North American Plant Protection Organization (NAPPO). 2005. Integrated pest risk management measures for the importation of plants for planting into NAPPO member countries. RSPM No. 24. Adopted October 2005. Ontario, Canada: North American Plant Protection Organization. http://www.nappo.org/Standards/NEW/RSPMNo.24-e.pdf.

USDA, Animal and Plant Health Inspection Service (APHIS). 2007. Importation of solid wood packing material: draft supplement to the final environmental impact statement. Washington, DC: U.S. Department of Agriculture, Animal and Plant Health Inspection Service. http://www.aphis.usda.gov/plant_health/ea/downloads/draft_swpm_seis.pdf.

USDA Foreign Agricultural Service. 2007. Global agricultural trade system. Washington, DC: U.S. Department of Agriculture, Foreign Agricultural Service. http://www.fas.usda.gov/gats/.

USDA Forest Service. 2004. National strategy and implementation plan for invasive species management. FS-805. Washington, DC: U.S. Department of Agriculture, Forest Service. http://www.fs.fed.us/foresthealth/publications/Final_National_Strategy_100804.pdf.

World Trade Organization (WTO). 1994. The WTO agreement on the application of sanitary and phytosanitary measures. Geneva, Switzerland: World Trade Organization. http://www.wto.org/english/tratop_e/sps_e/spsagr_e.htm.

Invasive Species and Disturbances: Current and Future Roles of Forest Service Research and Development

Mary Ellen Dix[1], Marilyn Buford[2], Jim Slavicek[3], Allen M. Solomon[4], and Susan G. Conard[5]

Abstract

The success of an invasive species is in large part due to favorable conditions resulting from the complex interactions among natural and anthropogenic factors such as native and nonnative pests, fires, droughts, hurricanes, wind storms, ice storms, climate warming, management practices, human travel, and trade. Reducing the negative effects of invasive species and other disturbances on our natural resources is a major priority. Meeting this goal will require an understanding of the complex interactions among disturbances, development of tools to minimize new invasions, and effective management of systems that have already been changed by invasive species.

In this paper, we suggest desired resource outcomes; we offer considerations for developing management strategies, policies, and practices needed to achieve these outcomes; and we note potential interactions of invasive species with other disturbances. We then identify invasive species-related research and development actions needed to achieve the desired outcomes. Interacting factors that influence desired outcomes include weather conditions, fire, pests, land use decisions, transportation, human health, human travel, and potential deployment of genetically engineered plants and animals. Disturbance and its interactions with invasive species can have ecological, social, and/or economic effects.

Forest Service Research and Development (R&D) priorities should focus on developing strategies, guidelines, and tools for mitigating invasive species and managing affected systems, as follows:

- Modeling the introduction and spread of invasive species to help proactively predict and prevent the introduction and establishment of an invasive species (also see prevention paper).

- Decision support, detection and monitoring tools and strategies for predicting, preventing, detecting, and responding to newly arrived invasive threats (also see prevention paper).

- Risk-cost-benefit analysis methodology to help determine the most effective management options.

- Strategies, systems, and practices for managing changed ecosystems to continue to deliver needed goods, services, and values.

- Tools that enable functional restoration of economically and/or ecologically critical systems.

- Strategies and guidelines to prevent, detect, monitor, and manage invasive species after major disturbances.

- Guidelines for economic, environmental, and social analysis.

Resources needed to accomplish the foregoing outcomes include the following:

1. Modelers skilled in multiobjective stand dynamics and forest management modeling.

2. Integrative specialists whose expertise incorporates ecological, social, and economic effects.

[1] National Program Leader for Forest Entomology Research, Forest Service, Research and Development, Washington Office, RPC-4, 1601 N. Kent St., Arlington, VA 22209.

[2] National Program Leader for Silviculture Research, Forest Service, Research and Development, Washington Office, RPC-4, 1601 N. Kent St., Arlington, VA 22209.

[3] Project Leader/Research Biologist, Forest Service, Northern Research Station, 359 Main Rd., Delaware, OH 43015.

[4] Retired; formerly National Program Leader for Global Change Research, Forest Service, Research and Development, Washington Office, RPC-4, 1601 N. Kent St., Arlington, VA 22209.

[5] Retired; formerly National Program Leader for Fire Ecology Research, Forest Service, Research and Development, Washington Office, 1601 N. Kent St., Arlington, VA 22209.

3. Functional specialists who provide data relating to basic processes and responses for use by integrative specialists and modelers.

4. Communication specialists skilled in print, Web, and novel technology transfer processes to provide support to functional and integrative specialists.

Introduction

The Forest Service and numerous other Federal, State, local, and private organizations recognize invasive species as a significant environmental and economic threat to the Nation's forests and rangelands. Interactions among invasive species and other environmental and anthropogenic disturbance regimes can exacerbate this threat (USDA Forest Service 2003). Actions taken to prevent, manage, and mitigate the adverse effects of invasive species and other threats depend on understanding the synergism among these disturbances and the potential effects of both disturbances and proposed mitigations on resources and on people's lives.

In this paper we suggest desired resource outcomes; considerations in developing management strategies, systems, policies, and practices needed to achieve these outcomes; and potential interactions of invasive species with other disturbances. We then identify invasive species-related research and development actions needed to achieve these outcomes. In the broad sense, our desired resource outcome is that forest and range ecosystems are healthy and productive and provide a sustainable supply of services, products, and experiences that enhance the

quality of life for present and future generations.

To meet this goal, we must consider the range and quantity of goods, services, and values that we will require our forests and rangelands to produce in the coming decades. Figures 1 and 2 show estimates of world and U.S. populations from 1950 through 2050. As populations and world economies continue to increase, so will the societal demands on our natural resources. We will rely on these lands to produce water, wood and non-wood products, recreational opportunities, biological diversity, and energy, all while playing a crucial role in climate change mitigation.

Although invasive species can have direct effects on many of these goods and services, it is important to recognize that these effects can also be greatly influenced by interactions with fire, weather and climate patterns, land use changes, and other disturbances. The influence of invasive species on critical natural resources may be increased or decreased in the context of other disturbances.

Future management, policy, and societal needs for research related to managing forests and rangelands under the influence of invasive species can largely be met through quantifying and projecting system behavior and value under different scenarios. At varying time and space scales, these needs include probabilistic projections of the magnitude and direction of change; likely outcomes without intervention; options for management actions, including their costs; and systems and practices for accomplishing these actions. Critical research deliverables include methods and tools for cost-benefit-risk

Figure 1.—*World population and estimates, 1950–2050 (United Nations 2007).*

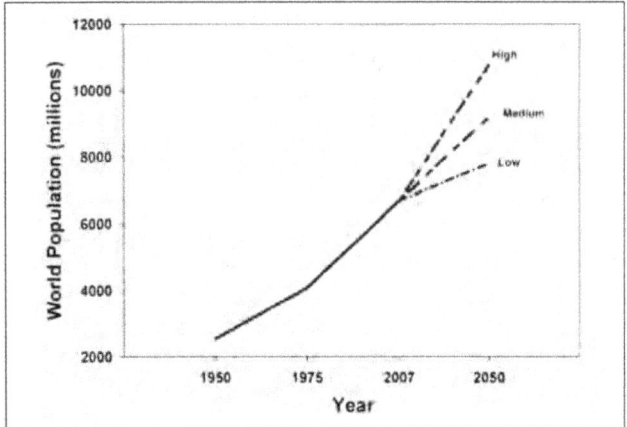

Figure 2.—*U.S. population and estimates, 1950–2050 (U.S. Census Bureau 2004).*

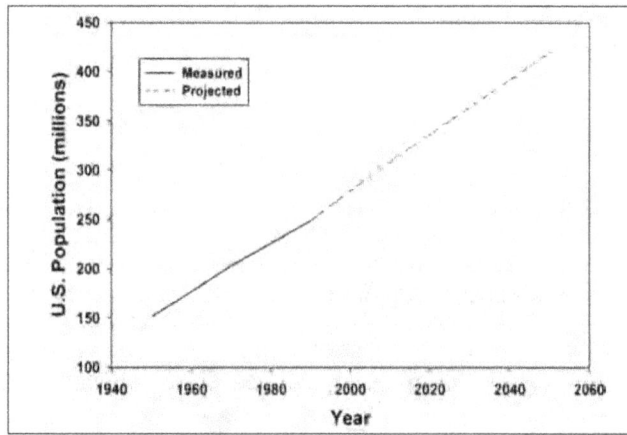

analysis for invasive species management options; collection of the basic data needed to populate, parameterize, and develop these models; strategies, systems, and practices for managing changed systems to continue to deliver needed goods, services, and values; and the ability to articulate what changed systems can and cannot deliver in probabilistic terms.

Forest and Rangeland Disturbances

A diverse group of natural environmental disturbances has the potential to alter our Nation's forests and rangelands, including native pests, drought, fire, hurricanes, tornadoes, and ice storms. Also a number of anthropogenic disturbances can potentially affect natural resources; these disturbances include nonnative invasive species, international and regional trade, transportation, development, and fragmentation. These disturbances can occur individually but often come in combination, and interactions among these inherent and anthropogenic disturbances are not well understood. Despite our lack of understanding, we know these disturbances (and their interactions) can disrupt ecosystem functions, social benefits, and economies. The resulting effects can be severe and may cause significant lasting ecological and socioeconomic effects. Thus, understanding the effects of individual invasive species and their interactions with multiple factors will enable development of effective approaches for sustaining and enhancing ecosystem functions and resource benefits. Land managers and owners need effective strategies, tools, guidelines, and practices to anticipate disturbances, act to prevent or lessen their effects, and restore the function and productivity of affected ecosystems.

Forest Service R&D has unique opportunities and responsibilities relative to invasive species (USDA Forest Service 2008). The Forest Service is the only Federal agency that maintains a strong research focus on forest pests. Although future trends are not certain, there is consensus that a small fraction of nonnative invasive species will interact with other anthropogenic and natural disturbances to disrupt existing ecosystem functions and adversely affect the goods, services, and values expected from these systems. Also, there is consensus that another small fraction may interact to provide benefits to humankind. Managing these invasive species appropriately will pose an enormous challenge considering our limited understanding of their potential interactions with our wildland

ecosystems and with other disturbances. Such understanding is critical to sustainable management of the Nation's natural resources. Our challenge over the next 50 years is to enhance our ability to predict and monitor these phenomena sufficiently to develop effective strategies to productively manage systems increasingly affected by invasive species and to recognize and capitalize on any benefits. The situation is urgent because population increases, human movement, global trade, and climate change will continue to drive changes in the world's biota—increasing the number and complexity of invasive species and disturbance interactions.

Key Disturbances

Our future success in both preventing introduction and establishment and managing spread of invasive species depends on our understanding of the interactions of diverse environments and disturbances as they impact invasive species ecology, and on our ability to manage ecosystems to minimize adverse outcomes. Key disturbances that can impact invasive species success include: severe or changing weather and climate conditions; fire regimes and their management; insect pests and diseases; land use and land cover changes; trade and transportation within and across regions; human health and travel; management practices designed to mitigate disturbance effects; and genetically engineered plants and animals. These types of disturbances are discussed below.

Severe or Changing Weather and Climate Conditions

Climate is clearly changing and, with it, the extremes of weather and climate as reflected in storminess, heat waves, minimum temperatures, droughts, and floods. In addition to chronic warming, multiyear and multidecadal climate cycles are being increasingly documented (Sutton and Hodson 2005). The familiar El Niño-Southern Oscillation (ENSO) brings increased rainfall to the Southwest and South and decreased rainfall in the Pacific Northwest and occurs for a year or 2 every 3 to 7 years. ENSO alternates with neutral conditions or with La Niña, which results in an opposite pattern of NW-SW rainfall (Sutton and Hodson 2005). Cool, wet phases of the Pacific Decadal Oscillation (PDO) increase rainfall in the Pacific Northwest and decrease it in the South and Southwest (especially during La Niña conditions). The PDO shifts from one mode to the other (warm dry phase) at about 20-year intervals. When the warm dry phase of the PDO coincides with an ENSO event, the Pacific Northwest becomes even drier (Mote et al. 1999). Recently, a connection of the Atlantic Multidecadal Oscillation (AMO) with western North American climate at about 60-year intervals has been discovered (see description of precipitation effects, below). The warm phase of the AMO is associated with warmer, drier conditions throughout the Western United States (Kitzberger et al. 2007). Although these climate cycles shift the intensity and frequency of climate variations, they are superimposed on a steady trend of increasing global temperatures. They can temporarily ameliorate or amplify effects of warming by adding still more climate variations, but they do not change the upward direction of the warming trends.

Climate warming is the most important weather variable driving shifts in invasive species distributions (e.g., Logan et al. 2003). With some important exceptions, greenhouse gas-induced warming throughout the next few centuries is expected to be greatest at night, during winter, over land, and toward the poles. Scientists have already observed such patterns in North America, Europe, and Asia (IPCC 2007a). Hence, the most successful invasive species under climate warming are expected to be those that are currently limited by cold winters and cool spring nighttime temperatures. An example of effects of warmer low temperatures is the newfound ability of mountain pine bark beetles to mature twice as fast as they used to, completing two life cycles a year in the Southwestern United States and at least one complete life cycle every year in areas of central British Columbia (Carroll et al. 2004). Warming nighttime temperatures are also responsible for beetle migration to higher elevations in the intermountain West, causing considerable mortality in limber pine populations that previously had not been subject to beetle attacks (Hicke et al. 2006; Logan et al. 2001, 2003).

Precipitation is increasing in some areas, and this trend is expected to continue because climate warming increases the intensity of the hydrological cycle, leading to greater evaporation and evapotranspiration, greater return rainfall, and greater runoff. In higher latitude temperate regions, precipitation increases appear to result from greater frequency of intense storms. Increased hurricane intensity and more hurricane landfalls at more northerly locations are also expected, at least in the Atlantic (Emanuel 2005, Webster et al. 2005). The implications of increased storminess, runoff, and flooding are obvious in enhanced storm damage to trees and increased abundance of damaged trees that can serve as infestation loci for pests such as borers. Meanwhile, the increased areas of soil

disturbance from flooding and treefall will subject ecosystems to potential enhanced establishment of invasive plants such as tamarisk in the Southwest. Note that the foregoing illustrates the fact that the effects of climate change are often driven by the increased *climate variability and extremes* that overlie the slow, chronic increase in mean temperatures. This increase in climate variability is predicted to continue with warming and is likely to produce the most obvious effects on ecosystem functioning (Overpeck et al. 1990).

In other areas, greater intensity and frequency of drought conditions are expected, a trend perhaps already being reflected in the presence of chronic drought conditions in the Western United States (Breshears et al. 2005). One recent study concluded that current drought conditions in the West are likely to become the normal situation throughout the 21st century (Seager et al. 2007). Others suggest that current drought conditions in the West may be attributable to 40- to 60-year cycles of warmer and dryer conditions associated with the AMO, as inferred from tree-ring records documenting fire frequency (and therefore climate conditions) over the past several centuries (Kitzberger et al. 2007). Indeed, in lower latitude temperate regions, such as the Southern and Southwestern United States, the subtropical subsidence zone in which most global deserts are located is expected to extend northward during this century. This phenomenon is under way and has already been well documented (IPCC 2007a). One result of the interaction of drought with warmer temperatures is increased stress, which reduces tree and forest resistance to invasive pests and plants. Such a response is clearly evident in the piñon *Ips* beetle-induced dieback of piñon and ponderosa pine on the Colorado Plateau during the 2000–04 drought (Breshears et al. 2005).

Fire Regimes and Their Management

The patterns of wildland fire, both globally and in North America, have changed markedly over the past 20 to 30 years (IPCC 2007b, Westerling et al. 2006). Changes include increased severity of fires in many short-return interval forested systems and increased frequency of fires in many arid and semiarid shrubland systems. The annual average burned area in the United States has increased greatly in recent years (NIFC 2008). Annual burned area on Forest Service lands, for example, has averaged almost one million acres per year over the 20 years from 1987 through 2006. This is nearly four times the average annual burned area for the previous 50 years (1937 through 1986) (USDA Forest Service, 2007). About 1.4 million acres have burned per year on Forest Service lands since 1999 (NICC 2008), and the upward trend is continuing.

Fire interacts with the potential for invasive species in many ways. Removal of native herbaceous vegetation by frequent or high-severity fire, intensive grazing, or canopy closure due to fire exclusion may reduce the seed sources for native herbaceous species and make habitats available for invasive plant species such as introduced annual grasses. Even when fire regimes are restored, lack of native seed sources can make restoration difficult and may even encourage the persistence of invasive species (Brooks et al. 2004). Grazing animals may also provide a vector for disseminating those species on fur or through their droppings. A number of invasive grass species are believed to have been introduced from Spain in the early years of California settlement; many of these species spread rapidly, carried by humans or by being lodged in the fur of sheep and other animals (Bossard et al. 2000).

Fuel breaks and other fire suppression strategies and tactics have been found to promote the invasion of nonnative plant species in the Western United States (Keeley 2006). A specific study of fuel breaks in California found that nonnative plant cover was more than 200 percent higher in fuel breaks than in adjacent forests (Merriam et al. 2006). In addition, cover of nonnative species was greater in areas that had been subject to three or more fires than in those subject to only one fire. Currently no information is available on the role of fuel breaks in the invasion of nonnative species in the Southeast, but the long history of prescribed fire and the high frequency at which it must be applied suggest that nonnative invasive species could already pose extreme threats to forest diversity and management goals. Fuel breaks in the Southeast are roads that follow the perimeter of management compartments and ownership boundaries. Features such as these are ecologically similar to roads that promote invasion of nonnative species by increasing connectivity within the landscape (Foreman and Alexander 1998).

Invasive grass species (e.g., various bromes in the Great Plains; bufflegrass in the desert Southwest, non-native grasses in the East) or shrub and tree species (e.g., tamarisk in the West, honeysuckle and other shrub and vine species in the East) may increase fine fuel loads, fire hazard, and potential fire severity, and may also affect rangeland forage, wildlife habitat and other values. If these invasions lead to more frequent or more intense fires, native species may be further eliminated. Such changes can affect the ecological and resource values of plant communities and the habitat value for associated wildlife species (Brooks et al. 2004; Zouhar et al. 2008).

In other situations, invasive species may alter fuel structure or fuel moisture in ways that lead to decreased fire frequency, such as with the invasion of the succulent iceplant (*Carpbrotia* sp.) into coastal sage ecosystems in southern California (Brooks et al. 2004).

During periods of drought, tree and shrub species can become more susceptible to a number of native and nonnative insect pests and diseases and more susceptible to fire (Logan et al. 2001, 2003). Increases in insects and diseases may further add to the fire hazard in already-stressed stands.

The decrease of vigor and full or partial canopy mortality associated with invasive insect and disease pests may also lead to long-term increases in fuel loadings (e.g., as branches and boles fall to the ground as is occurring in Michigan and Ohio due to emerald ash borer mortality) (Carroll 2003). These high fuel loads have the potential to increase the intensity and severity of future fires, especially on dry sites where rates of decomposition of dead woody material are slow.

Climate models suggest that the trends toward warmer, drier climate and increases in prolonged drought that have occurred over the past 20 to 25 years in much of western North America are highly likely to continue for the foreseeable future. Climate data also indicate that similar trends are beginning to occur in the Northeast. These changes will increase the many stresses on ecosystems that make them susceptible to plant, insect, and disease invasions. We can expect longer fire seasons and increases in the number of large, high-severity fires (IPCC 2007b). We can expect more rapid population growth of insects whose life cycles and distributions are limited by low winter temperatures (e.g., western pine beetle). We can expect increased spread of invasive plant species, such as cheatgrass, that are native to warm, dry summer climates. And we can expect increased difficulty in restoring riparian systems and wetlands that have had their hydrology, vegetation, and fire regimes severely altered by introduced species such as tamarisk and giant reed (*Arundo donax*). Fire regimes may be drastically altered in some areas if invasive (native or introduced) insects or diseases cause permanent alterations in vegetation structure by elimination of key ecosystem dominants (Brooks et al. 2004).

Management can help reduce the effects of these complex interactions in the following ways (see, for example, Brooks et al. 2004; Zouhar et al. 2008):

- Reduction of hazardous fuel loads can reduce fire risk and severity and support the persistence of native species.

- Actively managing forest stands to reduce water and nutrient stress and increase tree vigor can increase resistance to drought-induced fire hazard and to insect and disease attack. Such changes may also affect patterns of snowmelt and seasonal hydrology in cold winter areas.

- Targeted control efforts with manual treatments, herbicides, fire, or biological control may eliminate local populations of invasive species that are causing alterations in fire regimes or other ecosystem properties. Intensive management action (planting, seeding, followup controls) may often be needed to restore desired species and habitats.

- Planting of species (or ecotypes) better adapted to current climatic conditions may decrease the likelihood of further spread of invasive species by increasing the ability of native species to outcompete them.

- Revegetation with species that are not susceptible to (or do not act as intermediate hosts for) certain insect and disease species can reduce their populations or slow their spread, reducing their impacts on fire susceptible vegetation.

- Active management of fire timing, frequency, extent, and severity (e.g., based on understanding of phenology and fire tolerance/susceptibility of both invasive species and other species) can help limit the spread and reproduction of some invasive species. For example, reintroduction of properly timed frequent surface fire in Eastern U.S. deciduous forests may reduce the vigor and limit the spread of invasive vines such as Japanese honeysuckle (*Lonicera japonica*).

Insect and Disease Pests

Recent regional invasions of nonnative species and outbreaks of native pests threaten the health of our forests, rangelands, and urban forests. Severe outbreaks of Sudden Oak Death disease (*Phytophthora ramorum*), emerald ash borer (*Agrilus planipennis*), hemlock woolly adelgid (*Adelgis tsugae*), gypsy moth (*Lymantria dispar*), white pine blister rust (*Cronartium ribicola*), southern pine beetle (*Dendroctonus frontalis*), western bark beetles (Dendroctonus spp.), and other pests have drastically changed ecosystem function, structure, and composition. These outbreak pests outcompete native species,

change the rates of ecosystem processes, alter food webs, and affect native wildlife. For example, by killing hemlock (*Tsuga canadensis*) in the East, hemlock woolly adelgid (*Adelgis tsugae*) also affects trout (Salmoninae) survival in mountain streams by altering stream shade and temperatures (Snyder et al. 2005). Chestnut blight (*Cryphonectria parasitica*) killed American chestnut (*Castanea dentata*) trees, once an abundant, fast-growing, highly valued tree species in Eastern forests that was a preferred food source for wild turkey and a wide variety of other grazers, provided high-quality wood, and was an important sink for carbon sequestration (Jacobs 2005). White pine blister rust (*Cronartium ribicola*) is destroying whitebark pine (*Pinus albicaulis*) in the mountains of the Western United States, an important food source for the endangered Clarke's nutcracker (*Nucifraga columbiana*) (Schoettle 2004). Emerald ash borer threatens ash (*Fraxinus* spp.) species throughout most of the United States. These ash species are widespread in natural forests and are also commonly planted in urban areas (Cappaert et al. 2005). The number and severity of pest outbreaks are increasing. This trend is expected to continue into the future (National Invasive Species Council 2008).

Climate change, increased droughts, more frequent storms, greater human activity, and other changes in disturbance regimes will fuel changes in ecosystem composition that will alter susceptibility to native and invasive pest species. Some pest species may benefit from these changes because of health declines among native organisms stressed by the changes in the conditions under which they evolved. Alien species or species that are not native to the ecosystem may be better adapted to establish and thrive in these changed abiotic and biotic conditions. Often these nonnative species are very adaptable and may work in concert with each other. For example, laurel wilt pathogen (*Raffaelea lauricola*) is carried by an invasive bark beetle (redbay ambrosia beetle (*Xyleborus glabratus*)), invasion by nonnative earthworms (Oligochaeta) may predispose an ecosystem to invasion by alien nonnative plants (or vice versa) (Heneghan et al. 2006), or invasion by one alien plant species may lead to subsequent invasion by a series of other alien plant species (D'Antonio; Meyerson 2002).

Land Use and Land Cover Changes

As world population increases, demand for increased food and fiber production will almost certainly adversely affect the survival of native species and enhance the distribution of invasive species. The number of forest landowners is rising but the average forest tract size is getting smaller, increasing the need for effective communication networks (Sampson and Coster 2000). Conversion of forests and rangelands to agricultural and urbanized lands can accelerate fragmentation of the landscape and inadvertently introduce invasive species. By 2050, approximately 8.1 percent of the coterminous United States is expected to be urban (Nowak and Walton 2005). These urban and urban-interface forest ecosystems will be expected to provide many of the same benefits as extant forest and rangeland ecosystems. Because of their linkages with trade and transportation hubs, however, these urban forests, rangelands and waterways will likely serve as initial invasion sites and as reservoirs of invasive species. Effective monitoring strategies and protocols for potential invasive species activity in these urban systems will play a key role in detection and management efforts.

Trade and Transportation Within and Across Regions

The number of unintentional introductions of invasive species into the United States has increased dramatically since the early 1960s, and it is likely that the rate of introductions will grow over the next several decades. A major factor in this increase has been the advent of the shipping container. These containers have facilitated development of highly automated systems to rapidly load and unload goods from ships and move them from ports to final destinations. The containerization of shipped goods has caused a significant decrease in shipping time and freight costs for many goods. Although enabling an unparalleled increase in global trade, containerization of goods has facilitated a large increase in invasive species introductions (Levinson 2006). These species arrive in dunnage, wood shipping crates and pallets, agricultural commodities, seeds, plant nursery products, pet and aquarium trade goods, ship ballast, etc. The rapid movement of shipping containers on trucks and railroad cars facilitates the movement of invasive species from ports to the rest of the country (Levinson 2006). Inland distribution centers being developed in Kansas City, MO; Columbus, OH; Tennessee, and other inland locations will likely become focal points for invasive species introductions in the future. The number of containers imported into the United States in 2005 was about 25 million (U.S. Department of Transportation 2006a; 2006b). It is estimated that container-shipping capacity will increase by 50 percent over the next 5 years. The accelerating

rate of global commerce will result in a continuing increase of invasive species introductions into the United States (Rich 2006).

Human Health and Human Travel

Invasive species can cause disruptions by both directly and indirectly affecting human health. For example, the browntail moth (*Euproctis chrysorrhoea*), a defoliator of a variety of deciduous trees and shrubs, causes dermatitis and respiratory problems when people come in contact with larval hairs. Indirect effects on people are occurring from unforeseen sources. Eleven people were killed in Bandon, OR, in 1936 by a fire propagated by a highly flammable invasive plant, gorse (*Ulex europaeus*), introduced from Europe (Simberloff 1996). Another example is the planting of Australian *Melaleuca*, Asian cogongrass (*Imperata cylindrical*), and Brazilian pepper (*Schinus terebinthifolius*) along roadsides in Florida. These plants have become costly hazards due to increased fires along roadways and are currently being removed at great expense (Simberloff 1996). Invasive species such as birds, rodents, and insects (e.g., mosquitoes, fleas, and lice) can serve as vectors of human disease. The Asian tiger mosquito (*Aedes* (*Stegomyia*) *albopictus*) can serve as an insect vector of disease. This mosquito has a broad host range that includes many mammals, birds, and reptiles and, consequently, can transfer diseases, such as West Nile virus, among many species, including humans (Laqnciotti 1999). In addition, people themselves can serve as vectors for invasive species and foreign disease after traveling outside the United States. The globalization of trade and associated increased business travel, coupled with an increase in leisure travel as a consequence of an increasing U.S. population, will likely continue to generate invasive species health effects on U.S. citizens.

Management Practices Designed To Mitigate Disturbance Effects

Forest Service R&D has a rich history of developing management practices to mitigate individual disturbances in many ecosystems. Often these practices focus on a response to a single disturbance and tend to target a single invasive species or taxon. For example, in the Pacific Northwest native plants are propagated for postfire rehabilitation and to minimize the establishment of invasive plants. Poplars and willows are planted along stream banks to mitigate the effect of floods. Chestnut and American elms are being bred to withstand pathogens. Small-scale field tests have been initiated to evaluate the effects of introduction of hybrid American elm into native ecosystems

(Eshita et al. 2003, Powell et al. 2005). Merkle et al. (2007) review transgenic tree genetics programs and the introduction of resistance into a tree (Merkle et al. 2007). The agency's Sudden Oak Death program represented a rapid, targeted response to an emerging issue (Rizzo et al. 2005). Insecticides and biological controls are used to manage gypsy moth and hemlock woolly adelgid outbreaks (Sharov et al. 2002, Solter et al. 2004, Ward et al. 2004). For the most part, these mitigation practices are not based on environmental, social, and economic cost analyses and are not applied in the context of a management system. Synthesis of this information is needed, and this may be the first step toward the development of risk-cost-benefit analyses.

Future management systems need to provide strategies, decision and implementation tools, and practices that can address multiple disturbances and their interactions over time. Environmental, social, and economic analyses of the benefits and costs will play an essential role in identifying which disturbances to address. These management systems need to be dynamic, be responsive to changing needs, and incorporate the leading edge of knowledge development.

Biological Control

Invasive species can establish and proliferate in their new habitat because they are separated from their coevolved natural enemies. Biological control (biocontrol), a long-term strategic management tool used to suppress target invasive species populations below an economically or ecologically relevant threshold, intentionally unites the target invasive species with their imported or native natural enemies. Biocontrol may be the only viable option for managing infestations occurring at landscape scales or in environmentally sensitive wildlands. Because of our international ties and national partnerships, the Forest Service is well positioned to build teams to conduct cooperative research on invasive species in their native habitats. These relationships provide opportunities for identifying emerging needs, tools for predicting and preventing introductions, and the control of species that have been introduced into the United States. Candidate biological control agents must be carefully studied to ensure that life-cycles and behaviors are matched with the phenological and ecological characteristics of the target invasive species. In addition, because of potential impacts on nontarget species, rigorous prerelease evaluations and long-term postrelease monitoring are needed to provide a

scientific assessment of agent safety and efficacy especially under fluctuating environmental conditions, including climate change.

Genetically Engineered Plants and Animals

A significant effort to develop genetically engineered organisms is being made by the corporate sector around the world. Most of this effort focuses on agricultural applications. A developing area is microbes that can enhance ethanol/biofuel production from cellulosic fiber and algae. Within the United States, the use of genetically engineered soybeans, corn, and cotton has become widespread. By 2005, herbicide-tolerant soybeans and cotton accounted for 87 and 60 percent of total soybean and cotton acreage, respectively. Insect-resistant cotton and corn comprised 52 and 35 percent of cotton and corn acreage, respectively, in 2005 (Fernandez-Cornejo et al. 2006). To date, the use of genetically engineered crops in agriculture has failed to generate any disturbances not already inherent in the practice of agriculture (Fernandez-Cornejo et al. 2006).

Several genetically engineered tree species have been developed to date and include hybrid poplar trees containing genes that confer greater tolerance to *Septoria musiva*, a fungal pathogen that limits the use of these trees throughout the Eastern United States (Liang et al. 2001). The only genetically engineered tree that APHIS has approved for commercial distribution is the papaya (*Carica papaya*). This release is limited to Hawaii and was done to prevent the loss of an entire industry from destruction by ringspot virus (*Potyvirus*). The requirement by APHIS for genetically engineered trees to be sterile has prevented the release of all but the papaya. Several approaches are being pursued to generate sterility; most affect some mechanism of flowering. It is possible that releasing tree species, engineered with pest resistance, to breed with the susceptible population of that species could be used to address invasive pathogens and insects. This approach, theoretically, could have addressed chestnut blight and Dutch elm disease and could still mitigate the impact of the emerald ash borer. To date, the use of genetically engineered organisms in the environment has yet to generate novel disturbances. Continued research to address potential unwanted effects coupled with a conservative position on the commercial release of genetically engineered organisms is likely to continue to prevent unwanted effects in the future.

Historically, genetic variation has been managed in various

ways, including seed movement guidelines, intensive and specialized breeding programs, and development of unimproved locally adapted, regionally appropriate seed sources. Although genetically appropriate material is called for in our native plant materials policy (FSM-2070) (USDA Forest Service), many species used in our restoration efforts lack suitable seed/propagation sources. In addition to developing locally adapted seed sources, we must also develop strategies, genotypes, and seed sources that will be adapted to both current and future conditions. This effort will involve deploying more genetically diverse populations and breeding for appropriate abiotic and biotic resistances. Successful deployment of the proper planting material requires that the infrastructure be in place to produce sufficient quantities of seed and seedlings.

Research and Development Priorities

Some of the highest priority research and development needs to effectively manage invasive species in the face of multiple interacting disturbances in a rapidly changing environment include:

- **Risk-cost-benefit analysis methodology to help determine the most effective management options.** Development of this methodology is critical to developing options for rational action, including their costs, and includes articulating what changed systems can and cannot deliver in probabilistic terms.

- **Strategies, systems, and practices for managing changed systems to continue to deliver needed goods, services, and values.** The ability to provide effective, responsive management systems rests on quantifying and projecting system behavior and value under different scenarios.

- **Develop tools that enable functional restoration of economically and/or ecologically critical systems.** Tools and guidelines that help identify rational actions based on the risk and cost-benefit analysis of prevention, detection, prediction, and management options are critical to managing disturbed and changing systems.

- **Strategies and guidelines to prevent, detect, predict, monitor, and manage invasive species after major disturbances.** Practitioners and governments have strategies and guidelines based on sound science for preventing, detecting, predicting, monitoring, and managing invasive

species in the wake of major disturbances.

- **Guidelines for economic, environmental, and social analysis.** These analysis tools and forecast maps will allow land managers and land owners to make better informed decisions about their prevention, monitoring, management, mitigation, restoration, and rehabilitation activities.

Literature Cited

Bossard, C.C.; Randall, J.M.; Hoshovsky, M.C. 2000. Invasive plants of California's wildlands. Berkeley, CA: University of California Press. 360 p.

Breshears, D.D.; Cobb, N.S.; Rich, P.M., et al. 2005. Regional vegetation die-off in response to global-change-type drought. Proceedings of the National Academy of Sciences. 102: 15144–15148.

Brooks, M.L.; D'Antonio, C.M.; Richardson, D.M., et al. 2004. Effects of invasive alien plants on fire regimes. Bioscience. 54(7): 677–688.

Cappeart, D; McCullough, D.G.; Poland, T.M.; Siegert, N.W. 2005. Emerald ash borer in North America: a research and regulatory challenge. American Entomologist. 51: 152–165.

Carroll, A.L.; Taylor, S.W.; Régnière, J.; Safranyik, L. 2004. Effects of climate change on range expansion by the mountain pine beetle in British Columbia. Pp. 223-232 In: Shore, T.L.; Brooks, J.E.; Stone, J.E., eds. Mountain pine beetle symposium: challenges and solutions. October 30–31, 2003, Kelowna, British Columbia. Natural Resources Canada, Canadian Forest Service, Pacific Forestry Centre, Information Report BC-X-399, Victoria, BC. 298 p.

Carroll, M. 2003. Testimony by Michael Carroll, Minnesota State Forester, on behalf of the National Association of State Foresters before the U.S. Senate Committee on Agriculture, Nutrition and Forestry, June 26, 2003, on H.R. 1904, the Healthy Forests Restoration Act of 2003. http://www.stateforesters.org/testimony/6.26.03.htm.

D'Antonio, C.; L.A. Meyerson. 2002. Exotic plant species as problems and solutions in ecological restoration: a synthesis.

Restoration Ecology. 10: 703–713.

Emanuel, K. 2005. Increasing destructiveness of tropical cyclones over the past 30 years. Nature. 436: 686–688.

Eshita, S.M.; Slavicek, J.M.; Kamalay, J.C. 2003. Generation of American elm trees with enhanced tolerance/resistance to Dutch elm disease through genetics. In: Proceeding of the 14th interagency research forum on gypsy moth and other invasive species. Gen. Tech. Rep. NE-315. Newtown Square, PA: U.S. Department of Agriculture, Forest Service, Northern Research Station. 20 p.

Fernandez-Cornejo, J.; Caswell, M., et al. 2006. The first decade of genetically engineered crops in the United States. USDA-ERS Economic Information Bulletin No. EIB-11. Washington, DC: U.S. Department of Agriculture, Educational Research Service. 36 p.

Foreman, R.T.T.; Alexander, L.E. 1998. Roads and their major ecological effects. Annual Review of Ecology and Systematics. 29: 207–231.

Heneghan, L.; Steffen, J.; Fagen, K. 2007. Interactions of an introduced shrub and introduced earthworms in an Illinois urban woodland: impact on leaf litter decomposition. Pedobiologia. 50: 543–551.

Hicke, J.A.; Logan, J.A.; Powell, J.; Ojima, D.S. 2006. Changing temperatures influence suitability for modeled mountain pine beetle (*Dendroctonus ponderosae*) outbreaks in the Western United States. Journal of Geophysical Research. 111: 1–12.

Intergovernmental Panel on Climate Change (IPCC). 2007a. Climate change 2007: the physical science basis. Contribution of Working Group I to the Forth Assessment Report of the IPCC. Geneva, Switzerland: IPCC Secretariat. 996 p.

IPCC. 2007b. Climate change 2007: impacts, adaptation and vulnerability. Contribution of Working Group II to the Fourth Assessment Report of the IPCC. Geneva, Switzerland: IPCC Secretariat. 976 p.

Jacobs, D.F. 2005. Evaluating the efficiency of carbon sequestration in American chestnut (*Castanea dentata*). Technical Update report # 1011518. Palo Alto, CA: Electric Power Research Institute. 32 p. http://www.nwtf.org/nwtf_newsroom/press_releases.php?id=12043.

Keeley, J.E. 2006. Fire management impacts on invasive plants in the Western United States. Conservation Biology. 20: 375–384.

Kitzberger, T.; Brown, P.M.; Heyerdahl, E.K., et al. 2007. Contingent Pacific-Atlantic Ocean influence on multicentury wildfire synchrony over western North America. Proceedings of the National Academies of Science. 104: 543–548.

Laqnciotti, R.S. 1999. Origin of the West Nile virus responsible for an outbreak of encephalitis in the Northeastern United States. Science. 286: 2333–2337.

Levinson, M. 2006. The box: how the shipping container made the world smaller and the world economy bigger. Princeton, NJ: Princeton University Press. 376 p.

Liang, H.; Maynard, C.A.; Allen, R.D.; Powell, W.A. 2001. Increased *Septoria musiva* resistance in transgenic hybrid poplar leaves expressing a wheat oxalate oxidase gene. Plant Molecular Biology. 45: 619–629.

Logan, J.A.; Powell, J.A. 2001. Ghost forests, global warming, and the mountain pine beetle (Coleoptera: Scolytidae). American Entomologist. 47(3): 160–172.

Logan, J.A.; Régnière, J.; Powell, R.A. 2003. Assessing the impacts of global warming on forest pest dynamics. Frontiers in Ecology and the Environment. 1: 130–137.

Merkle, S.A.; Andrade, G.M.; Nairn, C.J., et al. 2007. Restoration of threatened species: a noble cause for transgenic trees. Tree Genetics & Genomes. 3: 111–118.

Merriam, K.E.; Keeley, J.E.; Beyers, J.L. 2006. Fuel breaks affect nonnative species abundance in California plant communities. Ecological Applications. 16(2): 515–527.

Mote, P., et al. 1999. Impacts of climate variability and change: Pacific Northwest. A report of the Pacific Northwest Regional Assessment Group. Washington, DC: United States Global Change Research Program. 109 p.

National Interagency Coordination Center (NICC). 2008. 2008

Statistics and Summary. http://www.predictiveservices.nifc.gov/intelligence/2008_statssumm/charts_tables.pdf.

National Interagency Fire Center (NIFC). 2008. Historical data on wildland fire. http://www.nifc.gov/fire_info/fires_acres.htm.

National Invasive Species Council. 2008. 2008-2012 national invasive species management plan. Washington, DC: Department of the Interior, Office of the Secretary, National Invasive Species Council. 35 p. http://www.invasivespeciesinfo.gov/council/mp2008.pdf (15 December 2009).

Nowak, D.J.; J.T. Walton .2005. Projected urban growth (2000-2050) and its estimated impact on the US Forest Resource. Journal of Forestry. 103: 383–389.

Overpeck, J.T.; Rind, D.; Goldberg, R. 1990. Climate-induced changes in forest disturbance and vegetation. Nature. 343: 51–54.

Powell, W.A.; Merkle, S.A.; Liang, H.; Maynard, C.A. 2005. Blight resistance technology: transgenetic approaches. In: Steiner, K.C.; Carlson, J.E., eds. Proceedings of the conference on restoration of American chestnut to forest lands. Natural Resources Report NPS/NCR/CUE/NRR – 2006/001. Washington, DC: U.S. Department of the Interior, National Park Service, National Capital Region, Center for Urban Ecology. 79–86.

Rich, L. 2006. A sea change in ocean shipping. Economic Development America, Spring. pp. 17-20. http://www.eda.gov/PDF/EDAmericaSpring2006GlobalGateways.pdf.

Rizzo, D.M.; Garbelotto, M.; Hansen, E. 2005. *Phytophthora ramorum*: integrative research and management of an emerging pathogen in California and Oregon forests. Annual Review of Phytopathology. 43: 13.1–13.27.

Sampson, R.N.; L.A. Coster. 2000. Forest fragmentation implication for sustainable private forests. Journal of Forestry. 98(3): 4–8.

Schoettle, A.W. 2004. Developing proactive management options to sustain bristlecone and limber pine ecosystems in the presence of a non-native pathogen. In: Shepperd, W.D.; Eskew, L.G., comps. Silviculture in special places: proceedings of the National Silviculture Workshop. Sept 8–11, 2003. Proceedings

RMRS-P-34. U.S. Department of Agriculture, Forest Service, Rocky Mountain Research Station: 146–155.

Seager, R.; Ting, M.; Held, I., et al. 2007. Model projections of an imminent transition to a more arid climate in southwestern North America. Science. 316: 1181–1184.

Sharov, A.A.; Leonard, D.; Liebhold, A.M., et al. 2002. "Slow the spread"—a national program to contain the gypsy moth. Journal of Forestry. July/August: 30–35.

Simberloff, D. 1996. Impacts of introduced species in the United States. Consequences, 2(2):1-13. www.gcrio.org/CONSEQUENCES/vol2no2/article2.html.

Snyder, C.D.; Young, J.A.; Ross, R.M.; Smith, D.R. 2005. Long-term effects of hemlock forest decline on headwater stream communities. In: Onken, B.; Reardon, R., comps. Third symposium on hemlock woolly adelgid in the Eastern United States, Asheville, NC, February 1–3, 2005. FHTET-2005-01. Asheville, NC: U.S. Department of Agriculture, Forest Service: 42–55.

Solter, L.; D'Amico, V.; Goertz, D., et al. 2004. Research on microsporidia as potential classical and augmentative biological control agents of the gypsy moth. In: Proceedings, XV USDA interagency research forum on gypsy moth and other invasive species 2004. GTR-NE-332. 74–75. Newtown Square, PA: U.S. Department of Agriculture, Forest Service, Northern Research Station. 98 p.

Sutton, R.T.; Hodson, D.L.R. 2005. Atlantic Ocean forcing of North American and European summer climate. Science. 309: 115–118.

U.S. Census Bureau. 2004. U.S. interim projections by age, sex, race, and Hispanic origin. http://www.census.gov/ipc/www/usinterimproj/. Internet Release Date: March 18, 2004.

U.S. Department of Agriculture (USDA), Forest Service. 2003. Forest Service invasive species management and implementation plan. http://www.fs.fed.us/rangelands/ecology/invasives.shtml.

USDA Forest Service. 2007. Historical fire data on file with U.S. Forest Service Fire and Aviation Management. Washington, DC: U.S. Department of Agriculture, Forest Service.

USDA Forest Service. 2008a. USDA Forest Service strategic plan FY 2008–2012. http://www.fs.fed.us/publications/strategic/fs-sp-fy07-12.pdf.

USDA Forest Service. 2008b. 2070 Vegetation Ecology. Forest Service Manual 2070. https://fs.usda.gov/FSI_Directives/2070.doc.

U.S. Department of Transportation, Federal Highway Administration, Office of Freight Management and Operations, Freight Analysis Framework, 2006. http://www.ops.fhwa.dot.gov/freight/freight_analysis/nat_freight_stats/docs/06factsfigures/index.htm.

U.S. Department of Transportation, Maritime Administration. 2006. U.S. waterborne container trade by U.S. custom ports, 1997-2005, based on data provided by Port Import/Export Reporting Service. http://www.marad.dot.gov/MARAD_statistics/index.html as of April 27, 2006. http://www.ops.fhwa.dot.gov/freight/freight_analysis/nat_freight_stats/

docs/06factsfigures/fig2_9.htm.

United Nations. 2007. World population prospects: the 2006 revision. Highlights. New York: United Nations, United Nations Secretariat, Department of Economic and Social Affairs, Population Division. 119 p.

Ward, J.S.; Montgomery, M.E.; Cheah, C.A.S.-J., et al. 2004. Eastern hemlock forests: guidelines to minimize the impacts of hemlock woolly adelgid. NA-TP-03-04. Morgantown, WV: U.S. Department of Agriculture, Forest Service, Northeastern Area State and Private Forestry.

Webster, P.J.; Holland, G.J.; Curry, J.A.; Chang, H.-R. 2005. Changes in tropical cyclone number, duration, and intensity in a warming environment. Science. 309: 1844–1846.

Westerling, A.L.; Hidalgo, H.G.; Cayan, D.R.; Swetnam, T.W. 2006. Warming and earlier spring increase Western U.S. forest wildfire activity. Science. 313: 940–943.

Zouhar, K,; Smith, J. K.; Sutherland, S.; Brooks, M.L. 2008. Wildland fire in ecosystems: fire and nonnative invasive plants. RMRS-GTR-42-vol. 6. Ogden, UT: U.S. Department of Agriculture, Forest Service, Rocky Mountain Research Station. 355 p.

The Role of the Forest Service in the Economics of Invasive Species Research

Linda L. Langner[1], Jeffrey P. Prestemon[2], and
Thomas P. Holmes[2]

Introduction

Invasive species increasingly influence various sectors of the economy through their effects on agricultural, forest, range, aquatic, and urban ecosystems. Policymakers evaluating the actual and potential effects of invasive species are concerned with allocating scarce taxpayer resources among a variety of competing governmental actions. To make allocation choices, they need information about the costs and benefits of alternative policies. Unfortunately, little is known about the magnitude of economic damages caused by invasive species, the costs of alternative controls, or the underlying factors affecting invasion risks and spread rates, much less the effectiveness of money spent on invasive species management. Economists can provide a vision that synthesizes the connections among invasive species management options in ways that help decisionmakers. This comprehensive vision can improve the selection and targeting of resources to reduce economic, social, and ecological damages. This paper reviews past and current Forest Service research on the economics of invasive species, outlines specific research needs, and identifies possible emerging issues in invasive species economics and policy.

Roles of Forest Service Research and Development in the Economics of Invasive Species Research

The Forest Service, U.S. Department of Agriculture (USDA), is one of several Federal agencies that conduct or support research on the economics of invasive species. Other agencies include the USDA Economic Research Service (ERS), the U.S. Environmental Protection Agency (EPA), the USDA Animal and Plant Health Inspection Service (APHIS), and the U.S. Coast Guard. Each agency focuses its attention on invasive species issues tied to its mission. For example, ERS focuses primarily on the effects of invasive species on agricultural systems, and the EPA focuses primarily on the effects of invasive species on aquatic ecosystems. Forest Service economics research focuses on issues affecting the management of invasive species on forests and grasslands.

Forest Service Research and Development (R&D) has a 25-year history of research into economic aspects of forest insects and disease, which forms the backdrop for newer efforts to understand economic dimensions of invasive species. Past research often focused on native pests, whose effects became increasingly important as forest management intensified and plantation forestry became more widespread in the latter part of the 20th century. Two examples are research on fusiform rust (*Cronartium quercuum f.* sp. *fusiforme*) and the southern pine beetle (*Dendroctonus frontalis*). Fusiform rust is a widespread and damaging disease of loblolly pine (*Pinus taeda*) and slash pine (*P. elliottii*) in the Southeastern United States. A central objective of much of the economics research was to identify how landowners could lower their risks of infestations and damages from fusiform rust. Economists collaborated in a study with silviculturists and forest pathologists to quantify the net benefits of the use of rust-resistant seedlings in pine plantations. In terms of benefits and costs, the study found that the benefits of the research embodied in the development of the rust-resistant seedlings were 2 to 20 times greater than the cost of the research (Cubbage et al. 2000). One economic study used a timber supply and demand model to evaluate the short-run timber price and overall economic effects of a large-scale infestation of the southern pine beetle in Louisiana and Texas. Some of the techniques used in that analysis formed the basis for other studies examining the effects of various kinds of forest damage agents (e.g., Butry et al. 2001, Prestemon and Holmes 2000). The pine beetle study established that the net economic effect of a large-scale infestation could total into the hundreds of millions of dollars, and that wealth transfers between winners and losers emerge from such catastrophic events (Holmes 1991).

[1] Economist, Forest Service, Research and Development, Washington Office, RPC-4, 1601 North Kent Street, Arlington, VA 22209.

[2] Research Forester, Forest Service, Southern Research Station, 3041 Cornwallis Road, Research Triangle Park, NC 27709.

Invasive species have been the focus of more recent and ongoing economics research studies. The European subspecies of the gypsy moth (*Lymantria dispar*) has been in the United States for more than a century, but it was not until the latter half of the 20th century that economists took note of its apparent economic effects. The Slow the Spread (STS) Program, a Federal-State partnership designed to limit the spread of the gypsy moth, needed economic analysis to establish its economic efficacy. One Forest Service-funded study documented that the STS Program could generate 25-year benefits ranging from $0.8 to $3.8 billion, mainly derived from mitigating damages to residential landscapes (Leuschner et al. 1996, Sharov and Liebhold 1998, Sharov et al. 1998), which far exceed programmatic costs.

The international trade research into invasive species is focused on quantifying the overall timber market effects of a potential exotic invasion into the United States. Jointly funded by ERS, APHIS, and the Forest Service, the case study in that analysis focused on the Asian variety of the gypsy moth or its close relative, the nun moth (*L. monacha*). The published work quantifies how such an invasion into U.S. forests would affect all categories of forest products manufactured and traded by the United States. This long-run analysis was also able to quantify the potential effects of alternative intervention policies to limit the invasion risk by such moths (Li et al. 2007, Prestemon et al. 2006).

In addition to examining the commodity effects of forest invasive species, ongoing Forest Service research is investigating the nonmarket economic effects of exotic forest pests. Several nonnative invasive pests cause nonmarket economic losses that exceed the timber losses of affected species or are confined to noncommercial species. Failure to account for the nontimber economic effects therefore results in downwardly biased assessments of overall economic losses. In a northern New Jersey study area, for example, mortality of eastern hemlocks due to the hemlock woolly adelgid (*Adelges tsugae*) was estimated to reduce private property values by roughly $14,750 per acre (Holmes et al. 2006, Huggett et al. 2008), which greatly exceeds the comparable per acre timber value lost. In a similar fashion, ongoing Forest Service studies in California indicate that Sudden Oak Death disease (caused by *Phytophthora ramorum*) causes large economic losses to residential private property values, while the timber losses from this disease are minimal (Holmes and Smith 2008).

Private landowners can take protective measures to reduce damages from nonnative pests and are likely to do so if perceived economic benefits exceed the control costs. Notably, the benefits of forest pest control actions taken by one landowner are shared by other forest landowners within a community (i.e., they are public goods). Consequently, some members of the landowner community will not take these benefits into account and will fail to take protective action. Forest Service research has shown that landowners who are most likely to participate in nonnative forest pest control programs are aware of the effects of their protective measures on other members of the community and that economic surveys can be used to identify community members who are most (and least) likely to participate in forest pest control programs (Holmes et al. 2008).

The absence of credible, nationwide estimates of the costs and economic losses caused by nonindigenous forest pests and pathogens limits the ability of policymakers to evaluate tradeoffs between economic effects and potential policy measures targeted at reducing those effects. Consequently, theoretically and empirically rigorous analyses are being developed to provide a foundation for estimating the aggregate economic effects from forest invasive species (Holmes et al. 2009). The modeling process quantifies key market and nonmarket economic effects using site-specific microeconomic analysis. Then the dynamic evolution of economic effects across landscapes is modeled as a spatial-dynamic process. Finally, by viewing economic costs and losses resulting from current invasions—specifically, the gypsy moth, hemlock woolly adelgid, Sudden Oak Death disease, and emerald ash borer (*Agrilus planipennis*)—as a sample drawn from an underlying stochastic process, statistical methods are used to estimate the overall distribution of effects, from benign to catastrophic, and the expected losses from future invasions are computed. This comprehensive research program is being developed at the National Center for Ecological Analysis and Synthesis through the collaboration of forest economists, entomologists, and pathologists representing Forest Service R&D, international forest research agencies, academic institutions, and The Nature Conservancy.

In summary, the Forest Service has partnered with sister agencies within the Federal Government and others to carry out economics research into invasive species. In spite of its documented and potential benefits to both the agency and to society, economics research has received a small share of the invasive species funding within the agency. Economics research funding has been targeted at particular pests or narrow features of economic effects and tradeoff analyses rather than focusing on the overall science of the economics of invasive species. For example, to date, the agency has funded no large-scale effort to quantify the many economic effects of headline pests such as Dutch elm disease (*Ophiostoma novo-ulmi*) or the chestnut blight (*Cryphonectria parasitica*), although evidence suggests that their effects on Eastern forests may have been important (Schlarbaum et al. 1997). The research has not been sufficient to track the substitution effects of individual species losses on both market and nonmarket values in affected forests, which could lead to more scientifically sound policy decisions by lawmakers and land managers. Other new headline species, such as the emerald ash borer and the Asian longhorned beetle (*Anoplophora glabripennis*), have also escaped detailed economic analysis in terms of both their effects and their control efficacies. Furthermore, we are aware of no research that has been able to separately quantify the roles of international trade and international travel on risks of invasion and establishment of some potentially catastrophic pests in temperate and subtropical North American landscapes. Nor is there a thorough understanding of the roles of land use and economic activity on risks of invasive species introduction, establishment, and spread in the United States.

An economic approach to invasive species research provides several tangible benefits: (1) the economic effects of exotic species invasions on various economic sectors, including markets for goods (such as timber) and services, and on nonmarket values (such as aesthetics) can be identified and quantified; (2) the costs of invasive species control (including prevention, detection, slowing the spread, and eradication) can be evaluated in terms of their efficacy in reducing economic losses; (3) the broad-scale relationships between economic inputs and economic damages can be assessed using a suite of economic methods, including econometric analysis, mathematical models, and simulation studies; and (4) economic analyses can be framed in terms useful for policy analysis by identifying the economic tradeoffs inherent in a suite of alternative policies designed to prevent or control the invasion of exotic pests.

Historically, Forest Service research in the economics of invasive species has benefited from collaborations with other agencies. We anticipate that future research will be similarly structured, continuing to complement and enhance research

capacity both inside and outside the agency. Support for economic research on forest invasive species, however, has been quite limited; therefore, the opportunities for addressing the broader array of research questions are also limited. It makes sense, then, that economists familiar with the research describe the priorities to advance the discipline of the economics of invasive species.

Key Future Economics of Invasive Species Research Priorities

The lack of comprehensive economic assessments of the catastrophic damages caused by important invasive species on the production of market goods and nonmarket economic values, international trade, travel, and overall economic activity has meant that the Forest Service is unable to reasonably predict the long-run effects of current or future pests. In the following text, we describe research priorities that are essential in addressing these gaps in knowledge.

Priority 1: Optimal Allocation of Public Resources

Invasive species programs cover a range of management options, all of which compete for public resources. The tradeoffs between program costs and economic losses are often poorly understood. This lack of information makes it difficult to design programs that ensure the most effective use of public resources and virtually impossible to evaluate the efficiency of investment in these programs.

Allocation of scarce resources to prevention, detection, spread management, and eradication requires a framework that evaluates effects across the range of invasive species. Further complicating this process is the influence of other natural events (e.g., drought, fire) and human factors (e.g., trade, travel, land use change) on the introduction, viability, and spread of invasive species, as well as the effect of nonnative forest pests on wildfires and other disturbances. An integrated modeling framework is needed that links biological and economic models that address disturbance events and economic effects for comparing the benefits and costs of different allocation strategies. This type of modeling framework should be able to address temporal, spatial, and multiobjective goals. Key questions include the following:

- How should limited program resources be allocated among prevention, detection, and control, both offshore and domestically?
- What economic rules of thumb or formal tools efficiently allocate program resources among the following:
  - Alternative invasive pest guilds.
  - Pathways by which invasive species could enter the United States.
  - Commodities affected by or acting as conduits for invasive species entry.
  - Types of consequences (e.g., environmental and commercial).
- When should vulnerable imported commodities be banned or require particular control measures as a condition of entry?
- When and under what circumstances do economic considerations suggest that programs be terminated or shifted to an alternative program goal?

Priority 2: Incentives and Choices for Private Land Managers

Private landowners and land managers play an important role in invasive species management. Private owners manage most forests in the United States, and their choices can dramatically affect invasive species. Because private owners cannot capture economic returns from the public-good effects of some invasive species programs, they generally underinvest in invasive species management from the standpoint of social optimality. Various incentives, indemnity, or other compensation schemes are often needed to obtain cooperation from private landowners, whose actions affect the spread of an invasive species, or to compensate private entities for mandatory destruction of private (infested or diseased) property for the public good. Designing incentives or compensation programs and setting levels within those programs to obtain the desired behavioral response are inherently economic problems, often involving the potential for moral hazard or unintended consequences. Key questions in this area include the following:

- What are the economic implications of using insurance, regulation with indemnity, voluntary incentives, or other approaches to obtain needed behavioral responses?

- What concepts can guide the structure and level-setting within any or all such alternative approaches?
- How should assets be valued in indemnity or compensation schemes?
- How do nonmarket values get incorporated into appropriate incentives involving privately controlled resources?

A dynamic interchange also exists between agriculture and forestry. In much of the country, the owners of agricultural lands and forest lands are largely the same people. Land use changes between agriculture and forestry are driven partially by returns to investment in these alternative land uses. Agricultural policies, such as the Conservation Reserve Program, can alter incentives and affect the agriculture-forestry margin. It is unknown whether the presence of invasive species can alter land use decisions. Key questions include the following:

- Is changing land use an economically viable or rational response for landowners trying to minimize damages associated with invasive species?
- How are decisions about control methods affected when the landowner also has the option to change land use?
- Are landowner responses to invasive pests different for cropland and grazing land than for forest land?

Research could be undertaken to evaluate the potential for extending existing land use change models to incorporate how the presence of invasive species or perceived risk of invasive species affects land use decisions.

Priority 3: Integrated Risk Assessment and Forecasting

A major obstacle to the development of forest health protection programs, both within public agencies and with broad-based private landowner participation, is the prevalence of risk and uncertainty (e.g., Holmes et al. 2008). Although the risk associated with each stage of a biological invasion is rather low, the uncertainty associated with each risk estimate is quite large. Because the risk and consequences of a biological invasion can be influenced by management actions, and because the characteristics of an invasion might be of a kind not seen before, novel management approaches may be required. Although estimates of the average risk that an introduced species will become a pest can be computed using lists of introduced species for which their success or failure is known (e.g., Reichard and

Hamilton 1997), it is not known how well past invasions can realistically predict the risk of future invasions. Key questions are as follows:

- How should land managers make decisions about invasive species when scientific information about species introduction, spread, virulence, and damages is complex and incomplete?

- How can statistical or other models be used to update invasive species management strategies when new biological and economic information is revealed (and uncertainty is reduced)?

Priority 4: Public Awareness and Investments in Invasive Species Management

The risk and uncertainty associated with most biological invasions, combined with the public-good characteristics of invasive species programs, may help to explain why mitigation and adaptation strategies often lag far behind the initial arrival and establishment of invasive species. One key factor in developing a rapid response to invasive species is public participation (GAO 2005). This factor is especially important in the Eastern United States, where private forests dominate the forest landscape. Key questions include the following:

- What social and economic factors influence the likelihood that private forest landowners will take protective actions against invasive species?

- What are the risk preferences of private forest landowners regarding the threat of forest invasive species, and how do these preferences affect the likelihood of taking protective action?

- How do the forest protection investments made by some members of a community influence the likelihood that other community members will make forest protection investments?

- What are effective means of raising public awareness about the risks and consequences of a biological invasion?

Priority 5: Methods for Estimating Nonmarket Effects of Invasive Species

Market effects of invasive species are generally easier to measure than nonmarket effects. Data are usually more readily available for conducting market effect analyses, and more biological information is usually available about how invasive species affect timber volume, crop yield, and forage production than on nonmarket measures. Yet, the nonmarket effects of invasive species are especially important when invasive species affect recreation (e.g., fishing), wildlife, biodiversity, or residential and urban areas. Better measures of nonmarket effects would enhance our ability to assess the broad range of economic and social effects of invasive species on ecological functions, human uses, and local and regional economies.

A number of methods exist for evaluating nonmarket effects, including hedonic pricing, contingent valuation, travel-cost method, and conjoint analysis. Applications of these methods to invasive species are relatively rare. Key questions include the following:

- How can existing nonmarket valuation methods be used to reliably evaluate the overall magnitude of nonmarket economic effects of biological invasions? How large are the economic threats to nonmarket values relative to timber values?

- Which members of society bear the greatest losses in nonmarket values due to invasive species?

- How do alternative invasive species management approaches affect the nonmarket values of forests?

- Can benefit estimates from the nonmarket valuation literature be appropriately transferred to invasive species analyses?

- How can estimates of damages to nonmarket economic values be translated into more effective policy solutions?

- What are the most efficient policies for reducing the nonmarket economic risks and effects of nonindigenous forest pests and pathogens?

In addition to improving our ability to estimate nonmarket effects of invasive species and related management approaches, a framework for monitoring social and economic effects of invasive species over time would be useful. This framework could be used to connect to major management initiatives such as the National Fire Plan, the National Recreation Strategy, and the National Invasive Species Management Plan.

Priority 6: Evaluating Optimal Policies of Invasive Species Management With Climate Change

As Earth's climate changes, vegetation communities and disturbance rates are likely to change. With such changes come potential avenues for introduction and establishment of exotic plants and animals into the United States. Important economic aspects that need addressing include the following:

- How quickly are invasive species risks evolving, particularly of invasion into the United States by plants native to tropical and subtropical biomes, and how should trade and other phytosanitary policies change to limit the expected net effects of these changes?

- What role will altered rates and severities of natural forest disturbances play in invasive species risk, and what would be the best management practices to apply in response to affected public and private lands?

Priority 7: Relative Effects of Trade, Travel, and Economic Growth

Government policies and private land management regulations may need continual review and revision to respond to the effects of increased trade, travel, and economic activity in the United States and worldwide. Programs and policies should be periodically evaluated to ensure they minimize the net effects of expected rises in invasive species introduction, establishment, and spread. Important questions to address include the following:

- What are the relative roles of trade, travel, and economic activity in affecting invasive species introduction, establishment, spread, and economic effects?

- How could the United States Government develop policies that would balance invasive species risks and the costs of invasions affecting the freedom of movement of goods and people across borders?

Structuring Forest Service R&D for Effective Economics of Invasive Species Research

The organizational structure of Forest Service R&D is not a barrier to effective research on the economics of invasive species. Economists have successfully collaborated internally across units and research stations, with outside agencies, and with external cooperators to address common research interests. Despite this success, more could be done to encourage and facilitate improved cooperation between economists and biological scientists. Such collaborations could advance bioeconomic approaches to invasive species research, such as those pioneered by analysts such as Cubbage et al. (2000), Sharov and Liebhold (1998), and Sharov et al. (1998). Developing research projects across disciplines could yield research and technology transfer tools that have larger economic and societal benefits. Often, advances in economic research are limited by biological information.

Because Forest Service capacity for economics research is limited, cooperation within USDA and across Government agencies, with academia, and with other partners is critical to advancing the research agenda. Support for invasive species economics research has been quite limited in the Forest Service, and recent projects have often been funded from external sources.

Conclusions

Despite limited financial resources, economics research into invasive species has often been of high quality and impact, with published studies breaking new ground in economics and policy. Although international trade research has been more broadly focused on timber product market effects across multiple products and the Nation's set of international policies potentially affecting aggregate losses, much of the funded studies in invasive species have been more narrowly focused on particular pests. These efforts include those of current and urgent concern, including the hemlock woolly adelgid and Sudden Oak Death disease, and those of recent study, including the gypsy moth and some indigenous forest pests (southern pine beetle and fusiform rust). Much of this research successfully quantified the high net returns to Government research and invasive species management. It seems clear that additional efforts to quantify the effects and control costs, broad-scale factors involved in overall invasive species risks, and policy analysis will lead to additional successful efforts to document societal net benefits from research and management. With ongoing concerns of a changing climate and therefore a changing picture of invasive species risk and spread, in an era of limited Government spending, economic tools and economic perspectives will be needed to prioritize both how research money is spent and how managers and decisionmakers should allocate scarce resources to address invasive species concerns.

Literature Cited

Butry, D.T.; Mercer, D.E.; Prestemon, J.P., et al. 2001. What is the price of catastrophic wildfire? Journal of Forestry. 99(11): 9–17.

Cubbage, F.W.; Pye, J.M.; Holmes, T.P.; Wagner, J.E. 2000. An economic evaluation of fusiform rust protection research. Southern Journal of Applied Forestry. 24(2): 77–85.

Holmes, T.P. 1991. Price and welfare effects of catastrophic forest damage from southern pine beetle epidemics. Forest Science. 37(2): 500–516.

Holmes, T.P.; Aukema, J.E.; Von Holle, B., et al. 2009. Economic impacts of invasive species in forests: past, present, and future. In Ostfeld, R.S.; Schlesinger, W.H., eds. The year in ecology and conservation biology. Annals of the New York Academy of Sciences. 1162: 18–38.

Holmes, T.P.; Bell, K.P.; Byrne, B.; Wilson, J.S. 2008. Economic aspects of invasive forest pest management. In: Holmes, T.P.; Prestemon, J.P.; Abt, K.L., eds. The economics of forest disturbances: wildfires, storms, and invasive species. Dordrecht, The Netherlands: Springer: 381–406.

Holmes, T.P.; Murphy, E.A.; Bell, K.P. 2006. Exotic forest insects and residential property values. Agricultural and Resource Economics Review. 35(1): 155–166.

Holmes, T.P.; Smith, W. 2008. Linking sudden oak death with spatial economic value transfer. In: Frankel, S.J., Kliejunas; J.T.; Palmieri, K.M., tech. coords. Proceedings of the Sudden Oak Death third science symposium. GTR-PSW-214. Albany, CA: U.S. Department of Agriculture, Forest Service, Pacific Southwest Research Station: 289–298.

Huggett, R.J., Jr.; Murphy, E.A.; Holmes, T.P. 2008. Forest disturbance impacts on residential property values. In: Holmes, T.P.; Prestemon, J.P.; Abt, K.L., eds. The economics of forest disturbances: wildfires, storms, and invasive species. Dordrecht, The Netherlands: Springer: 209–228.

Leuschner, W.A.; Young, J.A.; Waldon, S.A.; Ravlin, F.W. 1996. Potential benefits of slowing the gypsy moth's spread. Southern Journal of Applied Forestry. 20(2): 65–73.

Li, R.; Buongiorno, J.; Zhu, S., et al. 2007. Potential economic impact of limiting the international trade of timber as a phytosanitary measure. International Forestry Review. 9(1): 514–525.

Prestemon, J.P.; Holmes. T.P. 2000. Timber price dynamics following a natural catastrophe. American Journal of Agricultural Economics. 82(1): 145–160.

Prestemon, J.P.; Zhu, S.; Turner, J.A., et al. 2006. The forest product trade impacts of an invasive species: modeling structure and intervention trade-offs. Agricultural and Resource Economics Review. 35(1): 64–79.

Reichard, S.H.; Hamilton, C.W. 1997. Predicting the invasions of woody plants into North America. Conservation Biology. 11(1): 193–203.

Schlarbaum, S.E.; Hebard, F.; Spaine, P.C.; Kamalay, J.C. 1997. Three American tragedies: chestnut blight, butternut canker, and Dutch elm disease. In: Britton, K.O., ed. Proceedings, exotic pests of Eastern forests; 1997 April 8–10; Nashville, TN. [Place of publication unknown]: Tennessee Exotic Pest Plant Council: 45–54.

Sharov, A.A.; Liebhold, A.M. 1998. Bioeconomics of managing the spread of exotic pest species with barrier zones. Ecological Applications. 8(3): 833–845.

Sharov, A.A.; Liebhold, A.M.; Roberts, E.A. 1998. Optimizing the use of barrier zones to slow the spread of gypsy moth (Lepidoptera: Lymantriidae) in North America. Journal of Economic Entomology. 91(1): 165–174.

United States General Accountability Office (GAO). 2005. Invasive species: cooperation and coordination are important for effective management of invasive weeds. GAO-05-185. Washington, DC: United States General Accountability Office. 93 p.

Effects of Nonindigenous Invasive Species on Water Quality and Quantity

Frank H. McCormick[1], Glen C. Contreras[2], and
Sherri L. Johnson[3]

Abstract

Physical and biological disruptions of aquatic systems caused by invasive species alter water quantity and water quality. Recent evidence suggests that water is a vector for the spread of Sudden Oak Death disease and Port-Orford-cedar root disease. Since the 1990s, the public has become increasingly aware of the presence of invasive species in the Nation's waters. Media reports about Asian carp, zebra mussels (*Dreissena polymorpha*), golden algae (*Prynmesium parvum*), cyanobacteria (*Anabaena* sp., *Aphanizomenon* sp., and *Microcystis* sp.), and New Zealand mud snail have raised public awareness about the economic and ecological costs of invasive species. Along with other Federal agencies, States, and communities, Forest Service R&D must work to fund the research needed to better understand the linkages of land and water as venues for ecosystem effects of invasive species biology. This paper will identify desired resource outcomes, address management strategies and systems needed to achieve the outcomes, discuss potential effects on riparian systems and water resources, and identify research and actions needed to achieve the desired outcomes.

Introduction

In 1993, the Office of Technology Assessment (OTA), a former office of the U.S. Congress, estimated that 10 percent of all nonindigenous species posed a threat to become a nuisance species. According to the U.S. Environmental Protection Agency (EPA), more than one-third of all States have waters that are listed for invasive species under section 303d of the Clean Water Act of 1977. Nonindigenous species cause ecological damage, human health risks, or economic losses. Invasive species are degrading a suite of ecosystem services that the national forests and grasslands provide, including recreational fishing, boating, and swimming; municipal, industrial, and agricultural water supply; and forest products. Degradation of these services results in direct economic losses, costs to replace the services, and control costs. Losses, damages, and control costs are estimated to exceed $178 billion annually (Daily et al. 2000a, 2000b). Although agriculture is the segment of the economy most affected ($71 billion per year), costs to other segments, such as tourism, fisheries, and water supply, total $67 billion per year. Although more difficult to quantify, losses of these ecosystem functions also reduce the quality of life. Such costs are not assessed as losses to ecosystem services.

A review of studies on the economic effect of invasive species in the United States found that most are of limited use for guiding decisionmakers who formulate Federal policies on prevention and control (GAO 2002). They focused narrowly on estimates of past damages to a few commercial activities, agricultural crop production, and accountings of the money spent to combat a particular invasive species. These estimates typically do not examine economic damage done to natural ecosystems, the expected costs and benefits of alternative control measures, or the possible effect on future invasions by other species. Initiatives by Federal agencies to integrate information on the likelihood of invasion, the likelihood of economic damage to commercial activities and natural ecosystems, and the likely effectiveness of control methods are hampered by a lack of necessary data and of targeted resources.

[1] Program Manager for Air, Water and Aquatic Environments, Forest Service, Rocky Mountain Research Station, 322 East Front St., Boise, ID 83702.
[2] Retired, Formerly National Program Leader for Fish and Aquatic Research, Forest Service, Research and Development, 1601 North Kent St., Arlington, VA 22209.
[3] Research Ecologist, Forest Service, Pacific Northwest Research Station, 3200 SW Jefferson Way, Corvallis, OR 97331.

The Forest Service has a major role in the management of invasive species. Forest Service Research and Development (R&D) has unique capabilities to address the complex interactions among natural processes, land use, water resources, and invasive species and to meet future challenges through collaboration across mixed ownership and agencies. The Forest Service plays a vital role in managing 192 million acres of 156 national forests and grasslands, including 2 million acres of lakes, ponds, and reservoirs; more than 200,000 miles of perennial streams; and more than 16,500 miles of coastline. The Forest Service also provides technical assistance for 731 million acres of forests, rangelands, and prairies managed by other Federal agencies, States, private owners, and tribes. The Forest Service faces future challenges to reduce the introduction and spread of aquatic and riparian invaders while helping to protect one of the Nation's most critical resources—drinking water. National forests and grasslands are the source of drinking water for 3,400 cities and towns, serving an aggregate population of more than 60 million people. More than 3,000 non-community water supplies, such as campgrounds, are also on National Forest System (NFS) lands. Public lands managed by the Forest Service and its cooperating agencies play a dominant role in the Western United States. Roughly 75 percent of all water originates on NFS lands, giving the Forest Service the primary influence on water resources. Water issues increasingly dominate the agency's interests, and understanding the major direct and indirect effects of invasive species on water quality, water availability, and aquatic biological integrity will be increasingly important to the Forest Service mission and strategic goals. Amid growing public and political concern, we may anticipate growing pressure from a suite of stakeholder groups demanding that the Forest Service take a more active role in researching and managing aquatic invasive species. These pressures will be complicated by opposing stakeholder values regarding control measures (e.g., use of pesticides and piscicides) and regarding the relative benefits or harm of particular exotic species (e.g., exotic fish and plants in Western streams).

Authorities

The Organic Administration Act of 1897, the Multiple-Use Sustained-Yield Act of 1960, and the National Forest Management Act of 1976 authorize the Forest Service to establish and administer national forests to secure the sustainable benefits of multiple uses for the American people. The Clean Water Act mandates that Federal agencies "restore and maintain the chemical, physical, and biological integrity of the Nation's waters" and ensure that actions they "authorize, fund, or carry out must not jeopardize the continued existence of any listed species or result in the destruction or adverse modification of critical habitat." The National Environmental Policy Act (NEPA) of 1969, Nonindigenous Aquatic Nuisance Species Prevention and Control Act of 1990, Lacey Act of 1900, and Endangered Species Act of 1973 authorize Federal agencies to prevent the introduction of invasive species; provide for their control; and take measures to minimize economic, ecological, and human health effects, including the effects of pesticides and biocontrol agents.

Building on these and other statutes, Executive Order 13112 (February 3, 1999; http://www.invasivespecies.gov) calls for Federal agencies to use relevant programs and authorities to:

1. Prevent the introduction of invasive species.

2. Detect and respond rapidly to, and control populations of, such species in a cost-effective and environmentally sound manner.

3. Monitor invasive species populations accurately and reliably.

4. Provide for restoration of native species and habitat conditions in ecosystems that have been invaded.

5. Conduct research on invasive species and develop technologies to prevent introduction and provide for environmentally sound control of invasive species.

6. Promote public education on invasive species and the means to address them.

In addition, Order 13112 states that Federal agencies shall "not authorize, fund, or carry out actions that it believes are likely to cause or promote the introduction or spread of invasive species."

The National Aquatic Invasive Species Act now before Congress would reauthorize and strengthen the National Invasive Species Act of 1996 to protect U.S. waters by preventing new introductions of aquatic invasive species. The legislation, which Senator Carl Levin, D-Michigan, is sponsoring along with Senator Susan Collins, R-Maine, would regulate ballast discharge from commercial vessels; prevent invasive species introductions from other pathways; support State management plans; screen live aquatic organisms entering the United States for the first time in trade; authorize rapid response funds; create education and outreach programs; conduct research on invasion pathways, and develop prevention and control technologies for those pathways; authorize funds for State and regional grants; and strengthen specific prevention efforts in the Great Lakes.

Research Needs Regarding Invasive Species Effects on Water Quality and Quantity

Forest Service R&D is providing research and science delivery leadership to integrate diverse objectives associated with the management and conservation of aquatic resources under threat from invasive species. To focus its research, Forest Service R&D needs to work with the NFS to identify the highest priority needs. Besides targeted invasive species research, much of the ongoing ecological research could incorporate invasive species as a stressor at little additional cost. Adams, et al. (this volume) discussed Forest Service R&D roles specific to aquatic invasive species research. This chapter focuses on the impacts of invasive species on water quality and quantity.

Key questions include:

1. What will we require our riparian systems and water resources to produce in the coming decades?

2. What are the major goods, services, and values that may be disrupted by invasive species?

3. How will invasive species affect water resources and what are the associated socioeconomic effects?

4. What are our future management, policy, and societal needs to mitigate or adapt to the effects of invasive species as they alter the ability of aquatic ecosystems to provide these goods, services, and values?

5. How can research provide management systems and strategies for interactions between invasive species and water resources to optimize continued future production of these goods and services?

6. What are the effects on native species biodiversity, and the noneconomic societal values for maintaining that biodiversity?

Water quality and quantity are affected by the plants and animals that live in or near aquatic environments, as well as by management actions taken to control these taxa. For example, stocked fish are capable of hybridizing with closely related native species (Fausch et al. 2006). Introduced hatchery rainbow trout hybridized with Alvord Redband Trout, resulting in its becoming extinct in the wild. The Red Shiner (*Cyprinella lutrensis*) is a highly competitive, aggressive invasive species that has been widely introduced in rivers and streams in the United States by bait bucket transfer and stocking as a forage fish (Nico and Fuller 2007; USDA Forest Service 2004). It is implicated in the decline of native fish through hybridization, competition, and the introduction of pathogens (Walters, et al. 2008; Deacon 1988; Holden and Stalnaker 1975; Mettee et al. 1996). Introduction of fish in aquatic systems can affect trophic relationships and set off "trophic cascades" with resulting declines in native species and degradation of water quality (Baxter et al. 2004; Eilers et al. 2007).

Whirling disease was first introduced to the United States from Europe in the early 1900s through infected brown trout that were brought to Pennsylvania. The highly infectious disease has gradually moved to lakes and streams in Western States. The disease can be spread by fish, people, dogs, birds, boat trailers, hip boots, and fishing equipment that have been in infected waters. After entering a water body, the virus persists through spores that can survive for up to 30 years, even in dried-up streambeds. In response to the spread of the disease, Colorado implemented a complicated and costly ($12M) management and remediation program that prevents the stocking of trout from hatcheries testing positive into waters where whirling disease has not been found. This prohibition includes wilderness areas and streams where native trout may be restored. Trout from positive hatcheries will be only stocked into waters where the parasite has already been found to minimize the risk of contaminating other watersheds; however, stocking infected fish only perpetuates the problem.

Research on the effects of fish stocking in native ecosystems may require incorporation of economic analyses of the tradeoffs between conservation of native species and reduced recreational enjoyment. This analysis would be especially complex in the Western United States where native and introduced species have both commercial and recreational value. Matthews et al. (2001) and Knapp (2005) reported reductions in native amphibian populations in montane lakes in the Sierra Nevada Mountains of California where trout had been stocked. Dunham et al. (2004) identified multiple negative associations of native species with introductions of nonnative trout in the Western United States. Arresting human-mediated transport of invasive species will require support for public education programs such as Stop Aquatic Hitchhikers (a partnership among several Federal and State agencies, private industry, and nongovernmental organizations). Such programs may be effective at reducing the inadvertent transport of aquatic invasive species and pathogens between water bodies, but they must be able to identify specific potential threats.

Because of accelerating invasion rates, widespread economic costs, and environmental damages caused by invasive species, colonization theory has lately become a matter of considerable interest to aquatic ecologists. A synthesis of models of population growth, invasion biology, and theoretical population biology might provide quantitative tools for risk assessment of biological invasions. Retrospective and predictive models derived from rangewide analyses of potential invasive species of concern could be used to map historic and current localities of species nationally and to conduct analyses of patterns of dispersal and future threats. In the face of climate change and shifting human populations, the most vulnerable water resources could be identified and neighboring population centers targeted for public education campaigns that would reduce the potential for recreational users to spread invasive species through inadvertent or intentional introductions.

The identification and risk assessment of potential biological invaders would provide valuable criteria for the allocation of resources toward the detection and control of invasion threats (Anderson et al. 2004; Orr 2003). Ricciardi and Rasmussen (1998) identified 17 species from the Black, Caspian, and Azov seas (the origins of the zebra and quagga mussels (*Dreissena* spp.) and round goby (*Neogobius melanostomus*)) that could invade the Great Lakes-St. Lawrence River system and other North American inland waterways via the same vectors as previous invasive species. Characterizing life history traits of invaders may require increased international collaboration to identify limiting factors in the native range of the alien species. Whittier et al. (2008) developed a calcium-based invasion risk assessment for zebra and quagga mussels (*Dreissena* spp.) for streams in the Western United States.

Monitoring and assessment need to be conducted at appropriate scales to identify associations between native and invasive species. This approach will also affect the ability to detect direct or indirect species interactions. Brown and Moyle (1997) suggest that the success or failure of introductions of stream species are a function of the ability of a species to survive the fluctuating, highly seasonal, flow regime. Vaughn and Spooner (2006) emphasize the importance of appropriate sample scale in examining potential associations between species. One reason for underlying disparate conclusions about the ability of native and invasive species to coexist may be the different spatial scales at which data have been collected. Studies of the effects of nonindigenous species on aquatic food webs may require retrospective studies of population abundance to identify the responses of native species to the arrival of an invasive species (Laxson et al. 2003). It is also important to improve our understanding of spatial patterns associated with invasions of nonindigenous species from patch to landscape scales.

Early detection and monitoring of invasive species require that the methods for species collection and identification are rapidly deployable, cost effective, applicable across a range of ecosystems, and capable of identification of multiple taxa. Traditionally, these approaches are based on morphology. Mass identification of multiple taxa, especially for diverse micro- and meio-faunal groups is time consuming, technically intensive, and costly. An emerging methodology for early detection and monitoring of invasive species is the use of "environmental DNA" or e-DNA (Darling and Blum 2007). This "DNA barcoding" or "community metagenomics" (Tringe and Rubin 2005) based on the limited persistence of DNA in the environment has been used to detect the presence of an invasive species in freshwater systems (Ficetola et al. 2008).

Scientists have demonstrated that the introduction of fish and other species into previously fishless systems has caused major changes in abundance and distribution of native amphibians, zooplankton, and benthic invertebrates, particularly in high mountain lakes. Scientists will need to continue to evaluate the effects of introduced fishes on invertebrates, amphibians, reptiles, birds, and bats. A topic area of interest is high mountain lakes where introductions of trout for recreational sportfishing have caused declines in native amphibian populations. Where climate change is affecting temperature regimes, increasing temperatures may favor the spread of invasive species or pathogens or increase competition among juvenile salmon for temperature refugia. Management activities that could reduce the water temperature to historic thermal regimes could reduce the effect of invasive species on native salmonids.

Insect and disease outbreaks often lead to increased harvesting of the host species, including preemptive cutting before the arrival of the damaging organism as well as post-mortality salvage logging. Although such harvesting is seldom included as an indirect effect of the outbreak, it often includes removal of non-host species and may generate more profound ecosystem disruption than the pest or pathogen itself (Foster and Orwig, 2006). Studies comparing changes in microenvironment, vegetation, and ecosystem processes initiated by infestation by HWA, salvage logging, and preemptive logging of hemlock indicate that logging initiated stronger ecosystem changes than HWA-induced mortality due to abrupt and larger micro-environmental and vegetation changes, soil scarification, and the presence of extensive slash. Dramatic alterations in nitrogen cycling followed harvesting and persisted for many years.

Consequently, preemptive cutting appears to pose the greatest threat for nitrate leaching, followed by logging of declining sites and then by decline in the absence of logging (Foster and Orwig, 2006). Compositional changes between harvested and infested stands were similar overall but occurred at very different temporal and spatial scales. Following logging, there was a much greater increase in shade-intolerant seedlings, saplings, and herb layers (Kizlinski et al. 2002). It is important to evaluate such effects more broadly to help land managers make informed decisions about the best response to invasive species and natural disturbances.

New assessment and monitoring tools are needed to help manage the diversity of aquatic species necessary for successful land management projects that will conserve and recover species at risk. Baseline scientific information collected before large catastrophic events helps set priorities and assists with resource restoration after events such as hurricanes and large-scale wildfires. For example, using fish and fish habitat surveys to map changes in species occurrences and abundances over time, managers can evaluate the effects of management treatments for invasive species on native and nonindigenous aquatic species. The information could be used to assess watershed condition and aquatic biodiversity for uses such as forest plan revisions, forest project planning,

NEPA evaluations, management indicator species status reports, populating the Natural Resource Information System database, State and sensitive forest species evaluations, analysis of potential effects to aquatic diversity of reservoir placement, and a year-long analysis of fish recovery from severe drought.

Scientists recognize that riparian communities are among the most susceptible to invasion by nonnative species. Nationwide, in many streams and rivers, the native plants and animals were adapted to a system of dynamic equilibrium that included flood disturbance and wildfire to maintain diverse structure, age class, and community composition. Today, dams, diversions, ground water extraction, channelization, grazing, roads, and recreational use have modified many of these streams and watersheds to create conditions that favor some of the most aggressive invasive species. Riparian forests support the highest density and diversity of breeding birds in the Desert Southwest. These forests were historically shaped by regular flood events that were instrumental in the recruitment of native vegetation. Native riparian vegetation provides numerous nesting sites for a variety of Southwestern birds and also supports specialized arthropods, such as cicada, which are an important part of birds' diets during the breeding season. Current research includes measuring vegetation and changes in structure and recovery of native and exotic vegetation at study sites and sampling arthropods, bird populations, and nesting success. Using this research, models can then be developed to propose management strategies aimed at mitigating the effects of altered disturbance regimes on riparian vegetation structure, arthropod abundance, and habitat utilization and nesting success.

The balsam wooly adelgid (*Adelges piceae*) has killed virtually all the adult firs in Great Smoky Mountains National Park, thereby eliminating almost three-fourths of the spruce-fir forests in the Southern United States. As this dominant canopy tree disappears, the forests become warmer and drier. The subsequent change in temperature jeopardizes the survival of several northern species that have persisted as Ice Age relicts in these cool, high-elevation Appalachian forests. Studies have documented that spatial patterns of insect damage were more severe along the stream courses and less severe away from the streams (Kimpel and Schuster 2002). Stand productivity and water use appear little affected until an intermediate threshold of damage has occurred. Enhanced soil moisture availability may first be noticed toward the end of the growing season. After trees reach heavily damaged status, water uptake and transpiration are severely reduced throughout the growing season, leaving substantially more water available for evaporation, runoff, and/or use by other plant species. HWA also exhibits clear spatial patterns of damage as it spreads through hemlock stands. Moderate to heavy damage is common 4 to 8 years after initial colonization. The greatest damage is found at presumed initial contact areas, but this pattern appears to fade over time. Managers can expect that mortality will first occur in these areas and that they may have more time to implement control strategies in more interior locations. Thus, managers can expect that both productivity and water use may be little affected until an intermediate threshold of damage has occurred.

Future Considerations

Although most studies of the effects of invasive species relate to direct effects on native species, indirect, synergistic, or cumulative effects on aquatic systems are less well understood. To manage the disturbance caused by invasive species, land management agencies will need to rely on interdisciplinary research that involves the skills of aquatic biologists, hydrologists, silviculturists, soil scientists, biogeochemists, pathologists, and others. Forest Service R&D can avail itself of core strengths in long-term research on its experimental forests and rangelands. Research can provide landscape-scale modeling to support early detection, risk assessment, and mitigation of the effects of management activities to reduce or eradicate invasive species. Improving survey and inventory monitoring designs to maximize the likelihood of early detection of invasive species based on predictive models of the intrinsic potential of stream corridors, lakes, and wetlands to harbor invasive species may help ensure that invaders do not gain a foothold in previously unaffected areas.

Invasive species affect the quantity and timing of runoff, erosion, sedimentation, and other natural physical processes and may affect water availability in general. The hydrologic effects of invasive riparian species, such as salt cedar (*Tamarix* spp.), which consumes 10 to 20 times the water used by native species, may lower the water table and dry stream reaches in some areas (Wiesenborn 1996). In other cases, invasive species may lead to increased susceptibility to flooding. Stands heavily damaged by HWA may experience increased soil moisture due

to reduced transpiration and may deliver increased runoff to headwater streams.

Various means exist for nonindigenous species to degrade water quality. Decreased flows reduce transport of nutrients. Increased runoff and erosion increase sedimentation or alter nutrient flux. Replacement of bunchgrass by knapweed has increased erosion, raising stream temperatures and reducing fish habitat (Lacey et al. 1989). Cheatgrass in the arid West shades out nitrogen-fixing soil crusts, decreasing nitrogen input to the ecosystem (Whisenant 1990). Decomposition of invasive plants, such as Eurasian water millfoil, alters the loading cycles of nitrogen and phosphorus. Following dieoff, bacterial decomposition of decaying plant material can reduce dissolved oxygen. Zebra mussels (*Dreissena polymorpha*) filter particles from the water column and concentrate nutrients in their feces, changing nutrient regime and enriching sediment. They also change water clarity and alter conditions for native species adapted to turbidity, were shown to accumulate and transfer water-borne contaminants to other benthic invertebrates (Bruner et al. 1994; Hart, et al. 2001), and contributed to a bloom of cyanobacteria in the Great Lakes (Vanderploeg et al. 2001). New Zealand mud snails composed such a major portion of the biomass in a Wyoming mountain stream that they consumed 75 percent of gross primary productivity, and their excretions accounted for two-thirds of ammonium demand (Hall et al. 2003). Tui chub were introduced in Diamond Lake, OR, in the 1950s as a forage fish. Tui chub eat microscopic zooplankton that would normally graze on phytoplankton in the lake. The decreased zooplankton population resulted in uncontrolled growth of several forms of algae, including the blue-green algae (*Anabaena* sp.), which released the toxin anatoxin-a into the water (Tanner et al. 2005). The effect of Tui chub is now more than the loss of fishing opportunity. It is affecting water quality and overall recreation use and is causing serious public health concerns (Eilers et al. 2007). Multiple efforts to eradicate the Tui Chub have failed to eliminate the species from the system.

Terrestrial and aquatic invasive species can dramatically alter the loadings of nutrients, clean sediments, and toxic pollutants into surface and estuarine waters. Invasive aquatic species, such as the zebra mussel, can alter the toxic effects and bioaccumulation of contaminants by altering pollutant fate and dynamics within water bodies (Endicott et al. 1998). Zebra mussels remove contaminant-bearing particles from the water column and deposit them in sediments. Contaminants become available to benthic invertebrates and enter the food web (Bruner et al. 1994). Among the more significant indirect effects is the increased pesticide exposure in the environment due to eradication efforts. The majority of pesticides is targeted for controlling exotic weeds, insects, and mites (Lee and Chapman 2001; Pimentel et al. 1992). As new exotic pests are introduced, the use of pesticides targeted for their control will increase with a commensurate increase in ecological and human health effects. In addition, by altering erosion, runoff, and deposition processes, terrestrial, wetland, and aquatic invasive species can substantially alter pollutant loadings into surface and estuarine waters.

The spread of exotic diseases such as West Nile Virus by water are related to the breakdown of the same ecological, social and economic barriers associated with the introduction of other nonindigenous species. Emerging infectious diseases are a key threat to conservation and public health, yet predicting and preventing their emergence is notoriously difficult. Recent evidence suggests that water is a vector for the spread of Sudden Oak Death disease and Port-Orford-cedar root disease. This finding poses a significant challenge for the future. Management options for these two water molds, including chlorination of firefighting water, are costly and are accompanied by their own environmental effects. Introductions of nonindigenous amphibians such as the bullfrog (*Rana catesbeiana*) may contribute to the spread of the chytrid fungus (http://www.werc.usgs.gov/fs/amphstat.pdf) that has been linked to severe population declines in native amphibians. Several of these exotic diseases have the potential to become serious regional or national public health threats, and their number and geographical extent are likely to increase with global climate change. To understand the cascade of ecosystem effects, research needs to identify the fundamental processes that prevail in undisturbed systems before we can evaluate detect the direct and indirect responses of those systems affected by invasive species.

Managers developing response plans to aquatic ecosystem disturbances need to understand the synergies and cumulative effects as well as the socioeconomic impacts in order to resolve conflicting values surrounding recreation or restoration activities, such as Burned Area Emergency Response restoration priorities, stream restoration, or fish stocking (Fausch et al. 2006).

Other research needs include:

- Develop genetics-based methods for early detection of invasive aquatic and terrestrial species by identifying likely invaders and susceptible habitats.

- Conduct research on geographic variation of invasive species to increase the understanding of the spread of these species after they are established.

- Determine invasive species effects across trophic levels (interactions of invasive plants and animals) in selected landscapes such as Hawaii, other Pacific Islands, and the Caribbean.

- Determine distribution and habitat relationships of introduced species in managed aquatic and terrestrial systems, and develop a risk analysis to determine priority species for research.

- Develop experimental approaches to investigating ecosystem responses to invasive removal in priority land types, such as wilderness.

- Develop methods for rapid detection and efficient monitoring of invasive species in large river systems or at large geographic scales.

- Determine whether salvage logging in riparian areas reduces the spread of invasive species, and what management options are available.

- Develop instructional materials for training field crews in the proper methods to decontaminate field equipment to prevent the spread of aquatic invasive species and pathogens.

Conclusions

According to the EPA, more than one-third of all the States have waters that are listed for invasive species under section 303d of the Clean Water Act. All waters currently listed as impaired by nonindigenous species have been determined by a case-by-case analysis. It is not clear if the Forest Service will ultimately consider invasive species such as noxious aquatic plants as pollutants for Clean Water Act purposes. To the extent that terrestrial and aquatic nonindigenous species affect watershed condition by altering erosion, runoff, and deposition processes, failure to account for these effects in total maximum daily load models could result in substantial errors in calculating load allocations.

Literature Cited

Andersen, M.C.; Adams, H.; Hope, B.; Powell, M. 2004. Risk Assessment for Invasive Species. Risk Analysis. 24: 787–793.

Baxter, C.V.; Fausch, K.D.; Murakami, M.; Chapman, P.L. 2004. Non-native stream fish invasion restructures stream and forest food webs by interrupting reciprocal prey subsidies. Ecology. 85: 2656–2663.

Brown, L.R.; Moyle, P.B. 1997. Invading species in the Eel River, California: successes, failures, and relationships with resident species. Environmental Biology of Fishes. 49: 271–291.

Bruner, K.A.; Fisher, S.W.; Landrum, P.F. 1994. The role of the zebra mussel, *Dreissena polymorpha*. In: Contaminant cycling: II. zebra mussel contaminant accumulation from algae and suspended particles, and transfer to the benthic invertebrate, *Gammarus fasciatus*. Journal of Great Lakes Research. 20: 735–750.

Daily, G.C.; Alexander, S.; Ehrlic, P.R.; et al. 1997. Ecosystem services: benefits supplied to human societies by natural ecosystems. Washington, DC: Ecological Society of America. Issues in Ecology. http://www.esa.org/science_resources/issues/FileEnglish/issue2.pdf.

Daily, G.C.; Soderqvist, T.; Aniyar, S., et al. 2000b. The value of nature and the nature of value. Science. 289: 395–396.

Darling, J.A.; Blum, M.J. 2007. DNA-based methods for monitoring invasive species: a review and prospectus. Biological Invasions. 9: 751–765.

Deacon, J.E. 1988. The endangered woundfin and water management in the Virgin River, Utah, Arizona, Nevada. Fisheries. 13(1): 18–24.

Dunham, J.B.; Pilliod, D.; Young, M.K. 2004. Assessing the consequences of nonnative trout in headwater ecosystems in Western North America. Fisheries. 29(6): 18–24.

Eilers, J.M.; Loomis, D.; St. Amand, A., et al. 2007. Biological effects of repeated fish introductions in a formerly fishless lake: Diamond Lake, Oregon, USA. Fundamental and Applied Limnology (Archiv fur Hydrobiologie). 169(4): 265–277.

Endicott, D.; Kreis, R.G., Jr.; Mackelburg, L.; Kandt, D. 1998. Modeling PCB bioaccumulation by the zebra mussel (*Dreissena polymorpha*) in Saginaw Bay, Lake Huron. Journal of Great Lakes Research. 24: 411–426.

Fausch, K.; Rieman, B.; Young, M.; Dunham, J. 2006. Strategies for conserving native salmonid populations from nonnative fish invasions. Gen. Tech. Rep. RMRS GTR-174. U.S. Department of Agriculture, Forest Service, Rocky Mountain Research Station. 44 p.

Ficetola, G.F.; Miaud, C.; Pompanon, F.; Taberlet, P. 2008. Species detection using environmental DNA from water samples. Biology Letters. 4: 423–425.

Foster, D.R.; Orwig, D.A. 2006. Preemptive and salvage harvesting of New England forests: When doing nothing is a viable alternative. Conservation Biology. 20: 959–970.

Hart, R.A.; Miller, A.C.; Davis, M. 2001. Emprically-derived survival rates of a native mussel, *Amblema plicata*, in the Mississippi and Otter Tail Rivers, Minnesota. American Midland Naturalist. 146: 254–263.

Holden P.B.; Stalnaker, C.B. 1975. Distribution and abundance of mainstem fishes of the upper and middle Colorado River basin. Transactions of the American Fisheries Society. 104: 217–231.

Jelks, H.; Walsh, S.; Burkhead, N., et al. 2008. Conservation status of imperiled North American freshwater and diadromous fishes. Fisheries. 33(8): 372–407.

Kizlinski, M.L.; Orwig, D.A.; Cobb, R.C.; Foster, D.R. 2002. Direct and indirect ecosystem consequences of an invasive pest on forests dominated by eastern hemlock. Journal of Biogeography. 29: 1489–1504.

Knapp, R.A. 2005. Effects of nonnative fish and habitat characteristics on lentic herpetofauna in Yosemite National Park, USA. Biological Conservation. 121: 265–279.

Lacey, J.R.; Marlow, C.B.; Lane, J.R. 1989. Influence of spotted knapweed (*Centaurea maculosa*) on surface runoff and sediment yield. Weed Technology. 3: 627–631.

Laxson, C.L.; McPhedran, K.N.; Makarewicz, J.C., et al. 2003. Effects of the non-indigenous cladoceran *Cercopagis pengoi* on the lower food web of Lake Ontario. Freshwater Biology. 48: 2094–2106.

Matthews, K.R.; Pope, K.L.; Preisler, H.K.; Knapp, R.A. 2001. Effects of nonnative trout on Pacific treefrogs (*Hyla regilla*) in the Sierra Nevada. Copeia. 2001: 1130–1137.

Mettee, M.F.; O'Neil, P.E.; Pierson, J.M. 1996. Fishes of Alabama and the Mobile Basin. Birmingham, AL: Oxmoor House. 820 p.

Nico, L.; Fuller, P. 2007. *Cyprinella lutrensis*. U.S. Geological Survey Nonindigenous Aquatic Species database, Gainesville, FL. Revision Date: April 11, 2006. http://nas.er.usgs.gov/queries/FactSheet.asp?speciesID=518. Accessed 11/1/2009.

Orr, R. 2003. Generic nonindigenous aquatic organisms risk analysis review process. In: Ruiz, G.M.; Carlton, J.T., eds. Invasive species: vectors and management strategies. Washington, DC: Island Press. 518 p.

Ricciardi, A.; Neves, R.J.; Rasmussen, J.B. 1998. Impending extinctions of North American freshwater mussels (*Unionoida*) following the zebra mussel (*Dreissena polymorpha*) invasion. Journal of Animal Ecology. 67: 613–619.

Tanner, D.Q.; Arnsberg, A.J.; Anderson, C.W.; Carpenter, K.D. 2005. Water quality and algal data for the North Umpqua River Basin, Oregon, 2005. Data Series 229. Washington, DC: U.S. Department of the Interior, U.S. Geological Survey. 11 p.

Tringe, S.G.; Rubin, E.M. 2005. Metagenomics: DNA sequencing of environmental samples. Nature Reviews Genetics. 6: 805–814.

U.S. Department of Agriculture (USDA), Forest Service. 2004. Biological assessment of the environmental impact statement for the revision of the Chattahoochee-Oconee National Forest Land and Resource Management Plan. http://www.fs.fed.us/conf/200401-plan/EIS_A-H.pdf. Link accessed 8/25/2009.

U.S. Government Accountability Office (GAO). 2002. Invasive species: clearer focus and greater commitment needed to effectively manage the problem. Report to executive agency officials. GAO-03-1. Washington, DC: U.S. Government Accountability Office.

Vaughn, C.C.; Spooner, D.E. 2006. Scale-dependent associations between native freshwater mussels and invasive *Corbicula*. Hydrobiologia. 568: 331–339.

Walters, D.M., Blum, M.J., Rashleigh, B.; [and others]. 2008. Red shiner invasion and hybridization with blacktail shiner in the upper Coosa River, USA. Biological Invasions. 10: 1229–1242.

Whisenant, S. 1990. Changing fire frequencies on Idaho's Snake River Plains: ecological and management implications. In: McArthur, E.D.; Romney, E.M.; Smith, S.D.; Tueller, P.T., eds. Proceedings, Symposium on Cheatgrass Invasion, Shrub Die-off and Other Aspects of Shrub Biology and Management. Las Vegas, NV. 1989. Gen. Tech. Rep., INT-2767. Ogden, UT: U.S. Department of Agriculture, Forest Service, Intermountain Research Station: 4–10.

Whittier, T.R.; Ringold, P.L.; Herlihy, A.T.; Pierson, S.M. 2008. A calcium-based invasion risk assessment for zebra and quagga mussels (*Dreissena* spp.). Frontiers in Ecology and the Environment. 6(4): 180–184.

Wiesenborn, W.D. 1996. Saltcedar impacts on salinity, water, fire frequency, and flooding. In: Proceedings of the Saltcedar Management Workshop. Holtville, CA: University of California Cooperative Extension, Imperial County: 9–12. http://www.invasivespeciesinfo.gov/docs/news/workshopJun96/Paper3.html.

Additional Reading

Hall, R.O., Jr.; Tank, J.L.; Dybdahl, M.F. 2003. Exotic snails dominate nitrogen and carbon cycling in a highly productive stream. Frontiers in Ecology and the Environment. 1(8): 407–411.

Kimple, A.; Schuster, W. 2002. Spatial patterns of HWA damage and impacts on tree physiology and water use in the Black Rock Forest, southern New York. In: Onken, B.; Reardon, R.; Lashomb, J., eds. Proceedings of the Hemlock Wooly Adelgid Symposium, Feb. 5–7, 2002. New Brunswick, NJ: New Jersey Agricultural Experiment Station and Rutgers University: 12.

Lassuy, D.R. 1995. Introduced species as a factor in extinction and endangerment of native fish species. American Fisheries Society Symposium. 15: 391–396.

Lee II, H.; Chapman, J.W. 2001. Nonindigenous Species—An Emerging Issue for the EPA Volume 2: A Landscape in Transition: Effects of Invasive Species on Ecosystems, Human Health, and EPA Goals. ORD/Region Nonindigenous Species Workshop, Washington, DC, on July 12–13, 2000. 53 p. http://www.epa.gov/owow/invasive_species/workshop/nisvol2.pdf.

Orwig, D.A.; Foster, D.R.; Mausel, D.L. 2002. Landscape patterns of hemlock decline in New England due to the introduced hemlock woolly adelgid. Journal of Biogeography. 29: 1475–1487.

Pimentel, D.; Acquay, H.; Biltonen, M.; et al. 1992. Environmental and Economic Costs of Pesticide Use. BioScience 42: 750–760.

Pimentel, D.; Lach, L.; Zuniga, R.; Morrison, D. 2000. Environmental and economic costs associated with non-indigenous species in the United States. BioScience. 50(1): 53–65.

Riley, L.A.; Dybdahl, M.F.; Hall, R.O. 2008. Invasive species impact: asymmetric interactions between invasive and endemic freshwater snails. Journal of the North American Benthological Society. 27: 509–520.

Schloesser, D.W.; Smithee, R.D.; Longton, G.D.; Kovalak W.P. 1997. Zebra mussel induced mortality of unionids in firm substrata of western Lake Erie and a habitat for survival. American Malacological Bulletin. 14: 67–74.

Strayer, D.L.; Caraco, N.F.; Cole, J.J., et al. 1999. Transformation of freshwater ecosystems by bivalves: a case study of zebra mussels in the Hudson River. BioScience. 49(1): 19–27.

Vanderploeg, H.A.; Liebig, J.R.; Carmichael, W.W., et al. 2001. Zebra mussel (*Dreissena polymorpha*) selective filtration promoted toxic *Microcystis* blooms in Saginaw Bay (Lake Huron) and Lake Erie. Canadian Journal of Fisheries and Aquatic Sciences. 58: 1208–1221.

Warren, M.L., Jr.; Burr, B.M.; Walsh, S.J., et al. 2000. Diversity, distribution, and conservation status of the native freshwater fishes of the Southern United States. Fisheries. 25(10): 7–31.

Wilcove, D.S.; Rothstein, D.; Dubow, J., et al. 1998. Quantifying threats to imperiled species in the United States: assessing the relative importance of habitat destruction, alien species, pollution, overexploitation, and disease. BioScience. 48: 607–615.

Changing Conditions and Changing Ecosystems: A Long-Term Regional and Transcontinental Research Approach on Invasive Species

Ariel E. Lugo[1] and Grizelle González[2]

Two conundrums: "*Biological invasions are natural and necessary for persistence of life on Earth, but some of the worse threats to biological diversity are from biological invasions. ... One can either preserve 'a natural' condition, or one can preserve natural processes, but not both.*" (Botkin 2001: 261)

Abstract

Emerging new ecosystems are products of human activity. They occur everywhere but particularly in degraded sites and abandoned managed lands. These ecosystems have new species combinations and dominance by invasive species and appear to be increasing in land cover. As new ecosystems emerge on landscapes, issues of social values and attitudes toward alien species and naturalness increase in relevance. Despite their ecological and socioeconomic importance, however, very little empirical information exists about the basic ecology and social relevance of these ecosystems. We propose regional and transcontinental ecological and socioecological research to address questions about the structure, functioning, and ecological services of new ecosystems.

Introduction

This paper addresses the subject of land use and land cover change in relation to invasive species and the emergence of new ecosystems. Its purpose is to propose the elements of a national invasive species research program that would anticipate issues and provide objective information to inform policy and management actions regarding the Nation's forests and rangelands.

The Challenge

Dramatic environmental change—some say unprecedented change—is anticipated in the United States. Such a level of environmental change is a challenge to land managers because it tests their capacity to react and adapt their conservation activities in the face of evolving conditions. These changing conditions involve changes in the gaseous composition of the atmosphere, air temperatures, rainfall patterns, land covers, disturbance regimes, and species composition. Unfortunately, we do not understand the effects of these expected changes on ecosystems, and the tendency is to concentrate management approaches on preventing change for fear of the ecological and economic consequences. This fear, however, leads our policies to the conundrums in the previous quotation. The challenge is compounded by the need to deal with invasive species while also keeping track of ecosystem-level changes caused by species invasions and their implications to the overall functioning of landscapes. The scope of a national invasive species research program must transcend the individual species and focus on the whole biotic spectrum, including community-, ecosystem- and landscape-level processes.

Future Environmental Setting

The Forest Service Resource Planning Assessment Report (USDA Forest Service 2006) anticipates a greater urban land cover, denser and increasingly urban human populations, greater exchange of species and connectivity between continents, and greater anthropogenic effects than exist today. Alien and native species are expected to become more invasive as community composition shifts away from the familiar ones of yesterday and today. Potential climate change and changing disturbance regimes add uncertainty to the mix. One critical

[1] Director, International Institute of Tropical Forestry, 1201 Calle Ceiba, Jardín Botánico Sur, Río Piedras, PR 00926.

[2] Research Ecologist, International Institute of Tropical Forestry, 1201 Calle Ceiba, Jardín Botánico Sur, Río Piedras, PR 00926.

aspect of these future environmental settings is the nature of the habitat modification by humans. Human activity can alter or disturb ecosystems in novel ways relative to natural disturbances. A clear example is the introduction of toxic substances to the environment. Subtler are land degradation activities over large areas, which preclude native species regeneration.

Emerging Patterns

The change in species composition of ecosystems influenced by human activity will be the most daunting ecological issue in the 21st century. Already this issue is dividing ecologists in a number of manifestations of the same phenomena (Ewel et al. 1999). For example, ecologists argue about whether we are at the onset of an extinction crisis of equal proportions as the ones caused by asteroid effects. Ecologists also argue about whether alien species cause extinctions and the discussion turns more controversial when the debate focuses on whether the presence of aliens "destroys," "disintegrates," or changes ecosystem functions and processes. Ecologists also cannot agree on whether the expansion of alien species homogenizes or diversifies landscapes. Finally, ecologists cannot predict or do not understand what makes some species become invasive, and they argue about whether these invasive species are the causes of environmental change or the reflections of environmental change.

While ecologists are developing consensus to these fundamental questions, they have no disagreement on the following important aspects of the unfolding scenario:

- The level and intensity of human activity is increasing.
- Land degradation continues.
- Species extinctions are occurring.
- Climate change appears more certain.
- Whole landscapes are changing.
- Many familiar ecosystems are not so familiar anymore.
- A larger number of alien and invasive species appear in most ecosystems.
- The cost of mitigating the effects of these species on agriculture and other land uses is increasing.
- Many alien species are becoming naturalized.

- The invasion and naturalization of alien species are leading to the formation of new species combinations (new ecosystems).
- Increased competition and symbiosis are emerging between native and alien fauna and flora.

All these trends have the same outcome: They result in changes in species composition of ecosystems. Clearly, as the 21st century unravels, changes in species composition of communities will become the dominant ecological phenomena that scientists and land managers must face. For the Forest Service to maintain a position of leadership in the 21st century, it must provide society with clear and reliable information for dealing with the changes in the biota that human activity is causing. At least two factors contribute to the confusion about invasive species issues: (1) lack of empirical understanding of the ecological changes taking place and (2) a clash of values in society. We discuss these two factors next.

Emerging Ecosystems

Ecologists lack empirical understanding of the phenomena that will drive 21st century ecology because, for decades, ecological attention was centered on balanced or steady-state native and mature ecosystems. Ecologists gave less attention to anthropogenic disturbances and modified ecosystems. As a result, the science of ecology has a lot of catching up to do in the area of acquiring empirical information about the new ecosystems on Earth. New, emerging, or novel ecosystems are those with novel combination of species and human agency; i.e., they are the result of deliberate or inadvertent human action (Hobbs et al. 2006, Lugo and Helmer 2004). These ecosystems emerge naturally in response to such human actions such as land degradation, enrichment of soil fertility, introduction of invasive species, and abandonment of lands previously managed. Hobbs et al. (2006) suggest that emerging new ecosystems developed in a zone between two land cover extremes: (1) "wild" and (2) intensively managed or even urban covers (fig. 1). In reality, novel ecosystems develop anywhere species invasions occur. Within this broad geographic expanse, ecosystems are subjected to a broad range of modification and recurrent disturbances that induce species succession and competition and the resulting modification of the original community.

Figure 1.—*The interface between land covers on which novel or emerging new ecosystems occur (details in Hobbs et al. 2006). Novel ecosystems arise from either the degradation and invasion of "wild" or natural/seminatural systems or from the abandonment of intensively managed systems. They also arise when alien species invade nondegraded or otherwise intact ecosystems.*

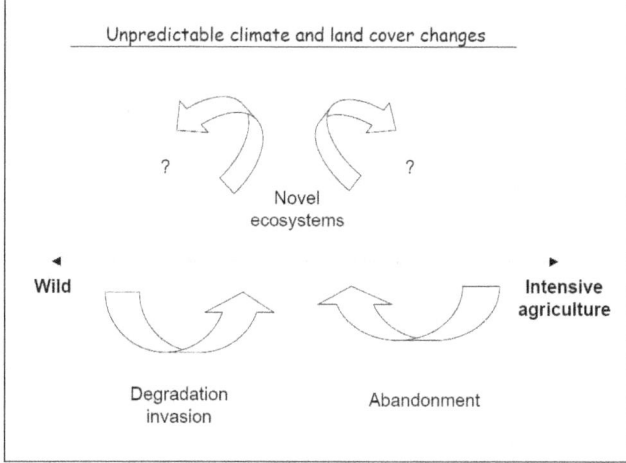

At least four processes leading to invasiveness and species composition change can be identified:

1. Severe alteration of site conditions because of land cover change or degradation can preclude historical succession patterns and favor species invasions.

2. Ecosystems are subject to invasion by alien species that exploit unused resources.

3. Invasive species can outcompete native species, even when the environment has not changed.

4. Each of these processes requires a different management strategy to mitigate it, because the ecological processes and consequences are different. Research is needed to unravel those differences and consequences.

Socioecology

A value issue is involved in how humans react to environmental change. We value familiar ecosystems and surroundings and naturally repudiate change, particularly when change is happening as fast as it is happening today. If we do not value an alien species, we are unlikely to accept information that suggests that that an alien species is acceptable, ecologically speaking. It appears that a need exists to combine empirical

science with social science to explore the role of values in the interpretation of empirical data and the use of social science principles to influence how empirical information is transmitted to people to make its use more effective. Although the need to merge ecological and sociologic knowledge is not new, the impetus for socioecology has gathered momentum. This momentum is fueled by expected global environmental change scenarios and the new emphasis on ecosystem services as a way of demonstrating the value of natural ecosystems to economic sustainability.

Research Approach

Research on invasive species must address multiple levels of biotic organization (from species to ecosystems and landscapes), be multidisciplinary, include a combination of natural history and experimental approaches, and encompass a long-term perspective over large spatial scales. We recommend a research program at two spatial scales to take advantage of the scope of the Forest Service Research and Development (R&D) organization and to address questions at the spatial level at which they occur. The two spatial scales are regional and transcontinental.

Regional Research

We suggest interdisciplinary long-term study of the intersection of values and empirical science in new ecosystems, such as plantation forests, urban rivers, polluted lakes and reservoirs, new forests on abandoned agricultural lands, invaded forests and rangelands, and vegetation on brownfields. Our suggestion involves collecting information on the functioning of these ecosystems and evaluating them empirically as to their role in the modern landscapes. Their functioning should be compared with familiar native ecosystems not invaded by alien species. Simultaneously, we suggest socioeconomic and anthropological studies to determine people's attitudes toward new ecosystems. Such studies must include people from many locations and professions, including ecologists. The objective is to uncover values at different geographic and intellectual distances from the location of ecosystems and ecological understanding, respectively. One aim is to find out to what degree science can inform society and modify its values. A hypothesis could be that informed science leads to informed decisions and modifies values. The questions embedded in figure 2 are as follows:

Q1a. To what degree are new ecosystems a product of new disturbance regimes?

Q1b. What is the relative importance of disturbance and the external genome in the formation of alien-dominated ecosystems?

Q2. What are the ecological characteristics of alien-dominated ecosystems?

Q3. What ecological services do alien-dominated ecosystems produce, and how do they compare to the native ones they replaced?

Q4. To what extent are the alien-dominated ecosystem services perceived, understood, and appreciated by humans, and how does human behavior respond to those services?

Q5. How do human outcomes and behavior alter the disturbance regime and to what degree are human attitudes toward alien species driving these changes?

The larger question is this: What determines dominance in ecosystems and how do ecosystem structure and process reflect that dominance across gradients (Denslow and Hughes 2004)?

Figure 2.—*A socioecological system with key research questions on invasive species. This diagram, which was developed as part of an exercise within the Luquillo LTER, relates the ecological drivers and consequences of alien species invasions (right) with the related socioecological drivers and consequences (left) and lists possible research questions.*

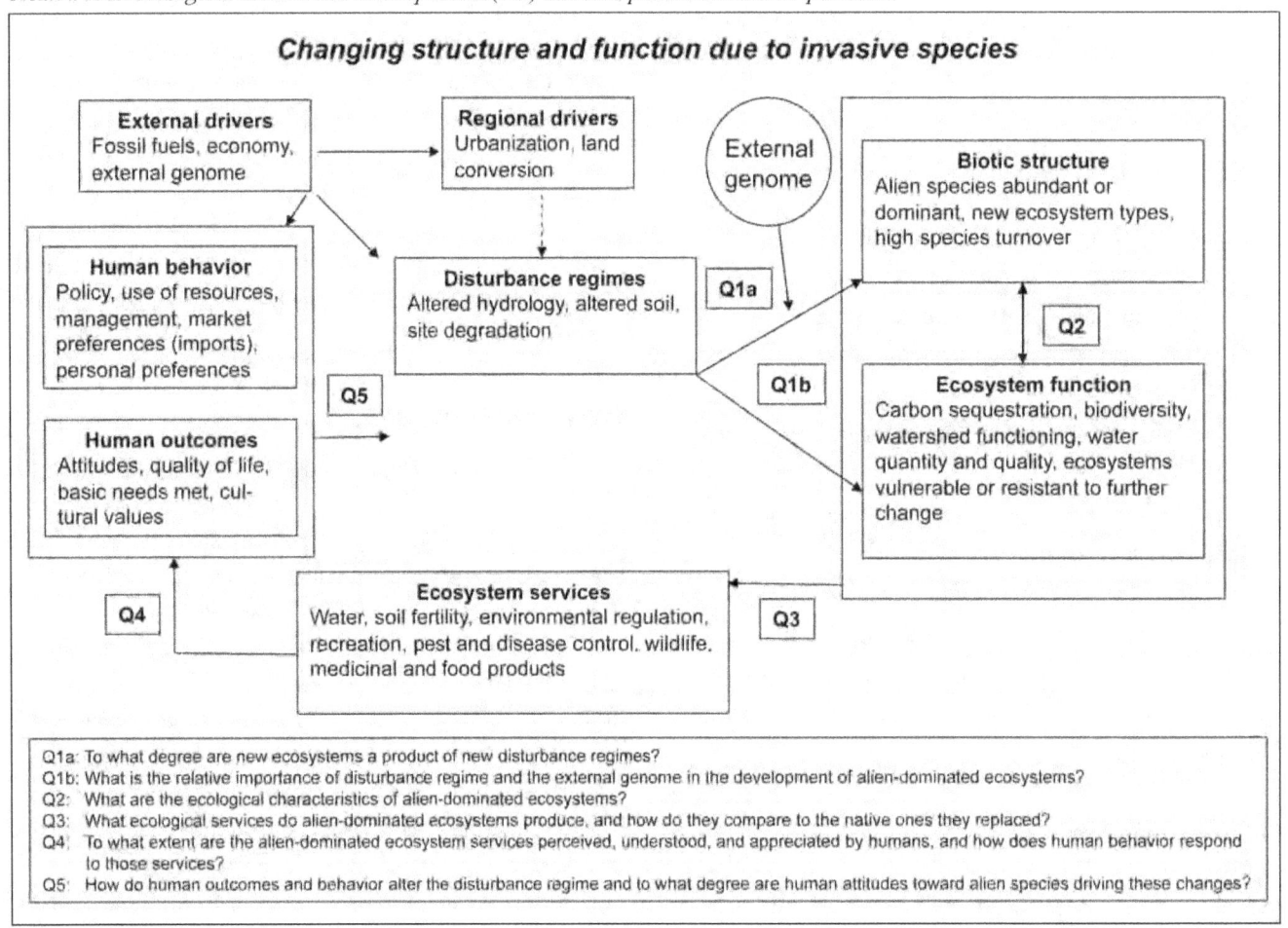

LTER = Long Term Ecological Research Network.

A Dynamic Invasive Species Research Vision: Opportunities and Priorities 2009–29

Transcontinental Research

Forest Service R&D units have access to many transcontinental gradients involving tropical to boreal (temperature), desert to rain forest (rainfall), lowland to montane (multiple gradients including atmospheric variation), etc. By focusing on gradients, researchers can quickly envelop the range of conditions that ecosystems face and can anticipate ecological behaviors at specific localities due to geographic shifts in the ecological gradients. Within these gradients (fig. 3a), they can find pristine to highly degraded examples of forests and ranges. Forest Service R&D is in a position to take advantage of the gradient space implicit in figure 3b to advance understanding of the mechanisms and consequences of emerging new ecosystems. We recommend a simple gradient approach to begin developing a research network with unified research questions. Whether the subject of research is at the species, group of species, or ecosystem level, studies across stations will benefit if they are conceived in the context of a gradient involving other stations or collaborators. Through commonality of methodology and types of questions asked, it will be possible to advance regional and transcontinental objectives simultaneously.

Some of the questions to address include the following:

- How do functional and structural attributes of emerging ecosystems vary along temperature, rainfall, and/or elevation gradients?

- How do ecosystem services change along temperature, rainfall, and/or elevation gradients?

- How does the dominance of invasive species change across these gradients?

- What are the patterns of plant life form dominance (e.g., trees vs. grasses) across emerging ecosystems and what are the implications to ecosystem functioning?

- Do plants, animals, and microbes follow the same patterns of response to human activity across the gradient?

Figure 3a.—*Temperature, precipitation gradients in watersheds within and without experimental forests and ranges. These gradients were developed as part of a NEON exercise among three Federal agencies: The USDA Forest Service, U.S. Geological Survey, and USDA Agriculture Research Service. Open dots illustrate sites that fall within steep temperature and precipitation gradients.*

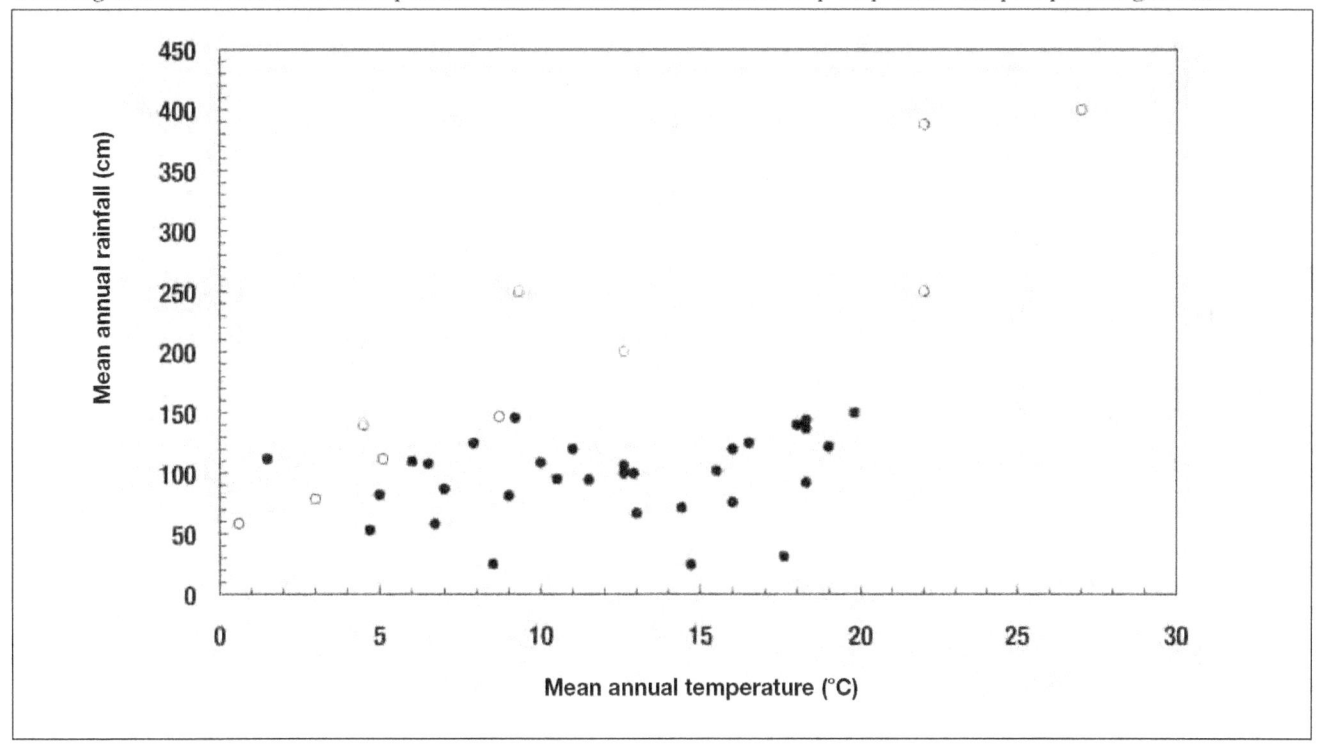

Figure 3b.—*Elevation gradients in watersheds within and without experimental forests and ranges. The insert in the elevation gradient represents the elevation gradient for the whole United States. These gradients were developed as part of a NEON exercise among three Federal agencies: the USDA Forest Service, U.S. Geological Survey, and USDA Agricultural Research Service. Open dots illustrate sites that fall within steep elevation gradients.*

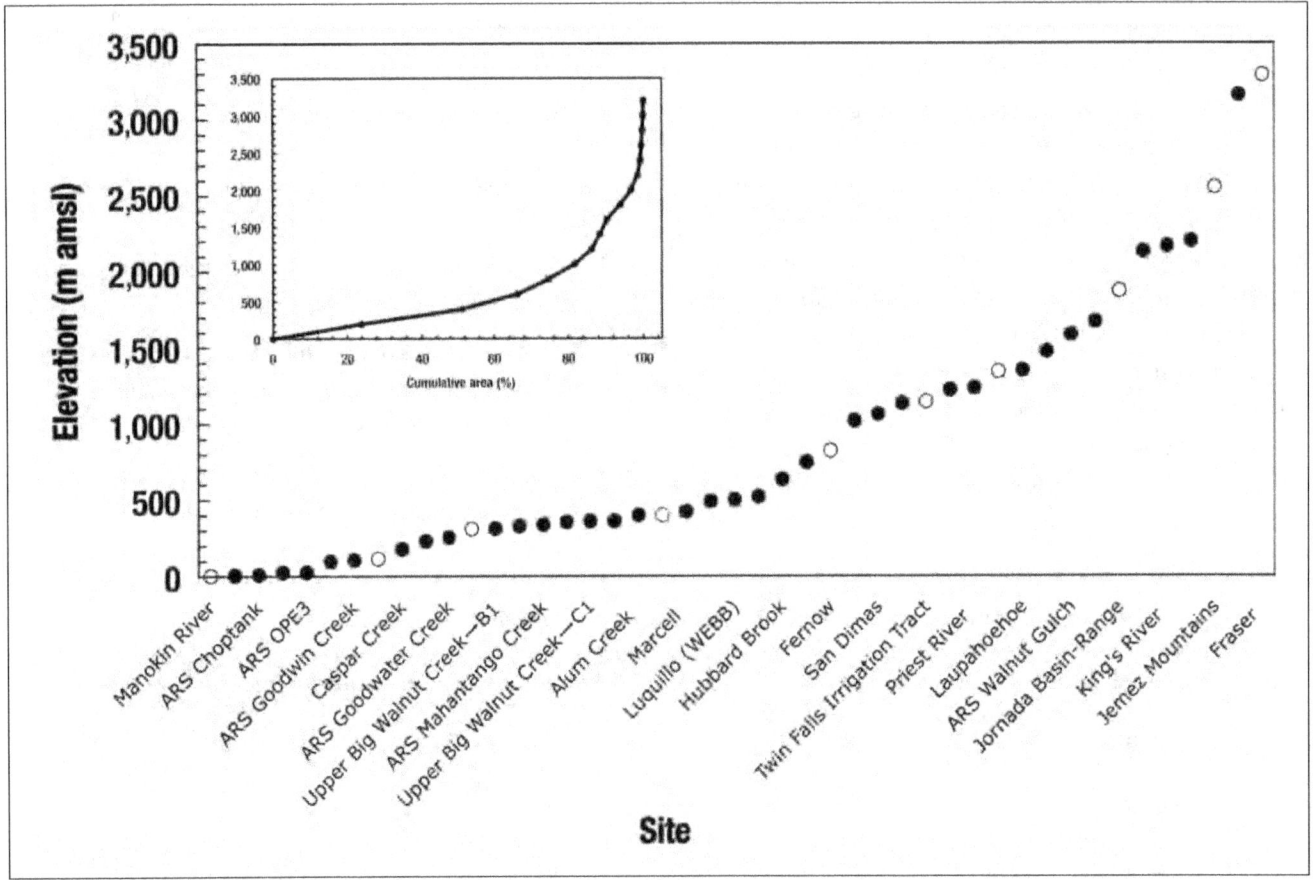

amsl = above mean sea level. ARS = Agricultural Research Service. NEON = National Ecological Observatory Network. USDA = U.S. Department of Agriculture. USGS = U.S. Geological Survey.

Acknowledgments

This paper was written in collaboration with the University of Puerto Rico. We thank Julie Denslow, an anonymous reviewer, and the editors for improving the manuscript.

Literature Cited

Botkin, D.B. 2001. The naturalness of biological invasions. Western North American Naturalist. 61: 261–266.

Denslow, J.S.; Hughes, R.F. 2004. Exotic plants as ecosystem dominants. Weed Technology. 18: 1283–1287.

Ewel, J.J.; O'Dowd, D.J.; Bergelson, J., et al. 1999. Deliberate introductions of species: research needs. BioScience. 49: 619–630.

Hobbs, R.J.; Arico, S.; Aronson, J., et al. 2006. Novel ecosystems: theoretical and management aspects of the new ecological world order. Global Ecology and Biogeography. 15: 1–7.

Lugo, A.E.; Helmer, E. 2004. Emerging forests on abandoned land: Puerto Rico's new forests. Forest Ecology and Management. 190: 145–161.

U.S. Department of Agriculture (USDA), Forest Service. 2006. Forest Service resource planning assessment report. Interim update of the 2000 RPA assessment. Washington, DC: U.S. Department of Agriculture, Forest Service.

The Use of Forest Service Experimental Forests and Ranges for Long-Term Research on Invasive Species

Ralph Holiday Crawford[1] and Gary W. Miller[2]

Abstract

The 81 experimental forests and ranges (EFRs) research sites make the U.S. Department of Agriculture (USDA), Forest Service unique among land management agencies. The EFRs were established for conducting applied research that serves as a basis for managing forests and rangelands. Most EFR research sites have long histories of experimentation and research that provide current and future answers to questions concerning the effects of management activities and how to better achieve management goals. Most EFRs have served as focal points for education and demonstration projects and as venues for the interaction between scientists and land managers and for the training of graduate students in forestry and related sciences. Research on EFRs has and will continue to contribute to a better fundamental understanding of how ecosystems function.

Introduction

Since 1903 the experimental forests and ranges (EFRs) have provided and continue to provide scientific information for managing national forests, rangelands and industrial and private lands (Adams et al. 2004). In accordance with Federal authority 4062.01 of the Forest Service Manual (FSM), section 4000, provisions of the Organic Administration Act of 1897 (16 USC 551), and the Forest and Rangeland Renewable Resources Research Act of 1978 (16 USC 1643), the Secretary of Agriculture has the authority to establish experimental forests and ranges. The major objective is that experimental forests and ranges will be used for conducting applied research that serves as a basis for managing forests and rangelands.

Forest Service Research and Development (R&D) has the primary role of conducting and administering research activities on EFRs. R&D structure includes the Pacific Northwest Research Station, Pacific Southwest Research Station, Rocky Mountain Research Station, Northern Research Station, Southern Research Station, International Institute of Tropical Forestry, and the Forest Products Laboratory. The number of experimental forests and ranges has exceeded 110 since 1903 but, with the processes of establishment and disestablishment, the current number is 81. For example, the Santa Rita Experimental Range was established in 1903 but has since been placed under the ownership of the State of Arizona. Some of the existing and oldest established permanent research sites and EFRs are the Fort Valley Experimental Forest (1908), Priest River Experimental Forest (1911), and the Bent Creek Experimental Forest (1927). The most recently established experimental forests are the Sagehen (2005), the Hawaii Experimental Tropical Forest (2007), and the Heen Latinee (2009). The distribution of EFRs covers vast geographical areas.

Experimental forests and ranges produce unique scientific products. Products have made significant impacts on the science community and have also affected policy at the national level. Table 1 lists a few examples of major scientific accomplishments at select EFRs in which research could focus on terrestrial and aquatic invasive species (plants and animals). EFRs that have long-term research focus on regeneration, harvesting, fire, climate change, and wildlife biology are well positioned for the study of invasion biology and competition between vegetation and other invasive species. Disturbances that contribute to the increase of invasive species on the landscape may be anthropogenic or natural and may interact in an integrated ecological process.

Most EFRs have uniquely valuable long-term studies and monitoring efforts that provide an invaluable record of recovery

---

[1] National Program Leader for Rangeland Ecology, Forest Service, Research and Development, Washington Office, RPC-4, 1601 N. Kent St., Arlington, VA 22209.

[2] Research Forester, Forest Service, Northern Research Station, Forestry Sciences Laboratory, 180 Canfield St., Morgantown, WV 26505.

Table 1.—*Experimental forests and ranges list of unique scientific products.*

Hubbard Brook Ecosystem Study	Major center for hydrologic research Research accomplishments on forest management Small watershed techniques for studying biogeochemistry First documentation of acid rain in North America Effects of forest-harvesting disturbance on water quality and quantity
Starkey Experimental Forest and Range	Pioneering sites that helped established the discipline of range management Wildlife habitat restoration and invasive vegetation Livestock grazing and wildlife effects on modification of mixed-conifer forests Invasive plants
Luquillo Experimental Forest	Tree species diversity and invasive plants Cross-site comparison of aquatic insect emergence ecosystems Canopy herbivory and soil processes in a temperate and tropical forest Earthworms (invasive species) and soil processes in tropical forest
Bent Creek Experimental Forest	Hardwood improvement cutting Long-term single-tree selection studies Ecological site-classification prediction models Intermediate stand management practices
Bartlett Experimental Forest	Habitat for wildlife Vegetation competition with regeneration methods
Hawaii Experimental Tropical Forest	Exotic grass cycles Nonnative terrestrial and aquatic plants and animals
San Joaquin Experimental Range	Development of sustainable grazing systems in oak woodland savannas Bird populations and diversity in oak woodland savannas
Fraser Experimental Forest	Subalpine forest ecology and hydrology affected by invasive beetles Invasive plants and animals in response to habitat changes

from disturbance and allow unusual events to be placed in the context of larger spatial and temporal patterns. EFRs address questions for forest and rangeland management at the appropriate scales of time and space; they are places to learn the fundamentals of natural ecosystem structure and dynamics. Finally, each EFR provides regional scientific information with well-documented disturbance and response histories, protected under special land use designations that allow manipulative research and protection of control sites (i.e., a secure research platform). There is an EFR in every ecoregion throughout the contiguous United States. The Forest Service has invested tens of millions of dollars into infrastructure, experiments, and long-term data collection and maintenance of EFRs. The activities of these resources are managed and maintained jointly by Forest Service R&D and the National Forest System (NFS). This work has long been recognized as having regional, national, and international importance as a strategy for research and science delivery for the 21st century (Crawford 2006; Miller and Crawford 2006).

This paper discusses the role of EFRs in future R&D efforts for invasive species research. The continued cutting-edge research activities and products from EFRs are the result of extremely strong partnerships between universities, other Federal agencies, tribal governments, State governmental agencies, private industry, and private landowners.

Critical Natural Resource Science Issues

The existing network of EFRs has a rich legacy of generating important research products and currently houses invaluable data sets that may be useful in addressing compelling natural resource science issues of interest to society. These issues generally involve considerable temporal or geographical scales that require pooling appropriate EFR data sets and integrating them with other national ecological networks to analyze regional or national problems. Examples of research problems that can be addressed by forming EFR networks include the following:

- The effect of global change on biodiversity, water yield, carbon sequestration, and ecosystem productivity.

- The causes and consequences of landscape changes such as fragmentation, urbanization, hydrologic alterations, and changing biotic patterns.

- The response of forest and rangeland ecosystems to disturbance, both natural and anthropogenic, to allow balanced resource utilization, habitat conservation, site restoration, and management of invasive species.

An important advantage of network-based research is that it enables the agency to address a common set of invasive species issues across different ecoregions. It facilitates measuring and interpreting results in a large-scale setting, thus giving resource managers research products of wider applicability on the landscape. For example, networking of research sites would shed light on the effects on both terrestrial and aquatic invasive species in ecosystems across regions, as opposed to an individual ecosystem.

It is important to develop the potential value of an EFR network while recognizing the continuing value of individual EFRs to address unique local resource issues. Defining large-scale science questions that require EFR partnerships may first be conceptualized according to strategic program areas (Fire, Resource Management and Use, Inventory Monitoring and Analysis, Invasive Species, Recreation, Water and Air, and Wildlife and Fish), thus allowing research priorities to be addressed from a national perspective across research work unit, programs, and station boundaries. In addition, individual EFRs have a tremendous potential to produce research products (data, information, knowledge, tools, and technologies) that are uniquely applicable to the ecosystem type in which they are located. Individual EFRs can continue to support sustainable land management within the local ecosystem types they represent, as well as collaborate with EFRs in other ecosystem types to produce research products that are applicable to much larger regional and national issues for invasive species.

Formation of High-Priority EFR Networks

Lugo et al. (2006) referred to the existing collection of EFRs as a network of permanent research sites, although it is understood that individual EFRs have unique attributes and data resources that may or may not be useful in studying certain research problems. In practice, data from subsets of EFR sites with the appropriate attributes can be integrated to address specific science questions at larger geographic scales. Similarly, subsets of EFRs may also be used to complement external networks (National Ecological Observatory Network, Long Term Ecological Research Network (LTER), Consortium of Universities for the Advancement of Hydrologic Science, Research Natural Area, Forest Inventory and Analysis, National Atmospheric Deposition Program, etc.) to further address large-scale invasive species science questions.

Emphasizing a network-based strategy can facilitate the integration of available knowledge, data sets, and technology to address regional and national environmental issues. Individual EFRs can continue to provide vital scientific discoveries applicable to local ecosystem conditions and land management challenges on public and private land. An integrated EFR network, however, also has the potential to yield vital scientific discoveries at much larger scales. For example, the LTER program supported by National Science Foundation is essentially a network of Forest Service experimental watersheds. A network of EFRs will allow for integration across broad temporal and geographical environmental gradients, thus generating new information in the expansive ecological space where complex invasive species management problems are emerging (Bailey 1991; Kneipp 2005; Lugo et al. 2006; Miller and Crawford 2006).

Summary

As we move forward in the 21st century, the experimental forests and ranges still possess a unique capability for answering large-scale invasive species questions of great societal benefit if they are integrated as a network of science assets. An integrated network of EFRs will enable resource managers and policymakers to address emerging local, national, and global natural resource issues pertinent to society.

A comprehensive strategy is proposed for updating the support, protection, management, and use of EFRs in the 21st century (Miller and Crawford 2006). The strategy will increase the agility of the EFRs and enhance their ability to respond to contemporary and emerging invasive species science issues. The prioritization of the initiatives is difficult, but the strategy is designed to be responsive to changing complex ecological and economic factors.

- Define the critical natural resource science issues on invasive species of interest to land managers and society that EFRs can address at regional and national scales.

- Support the formation of high-priority EFR networks and partnerships with land managers and other ecological networks to address critical invasive species science issues.

- Increase visibility of and support for EFR sites, infrastructure, and research programs by illustrating their value, relevance, and unique ability to address current and emerging invasive species issues of importance to the public, the agency, and the world science community.

- Enhance internal and external communication among Forest Service land managers and Research and Development scientists, as well as cooperators and customers associated with EFRs, so they can nurture relationships and cooperate in the research and science delivery process for invasive species.

- Develop flexible data networking tools and policies that integrate scientific information on flora, fauna, water, air, and soils, thus allowing EFRs to form effective internal and external networks across broad temporal and spatial scales.

When emerging research priorities are such that existing EFRs are not capable of providing appropriate scientific information, the Forest Service Chief has the authority under 7 Code of Federal Regulations (CFR) 2.60(a) and 36 CFR 251.23 to establish new EFRs as recommended by the station director with concurrence of the regional forester under FSM 4062.04. Additional EFRs may be needed to study 21st-century issues, such as invasive species associated with the urban-forest environments. If the EFRs are not integrated, the loss is essentially a costly, missed opportunity to conduct new science on immensely important research problems. Such a missed opportunity will delay our ability to build on 100 years of science on EFRs and reach the cutting edge of large-scale research efforts at a time when this type of effort is critically needed.

Literature Cited

Adams, M.B.; Loughry, L.; Plaugher, L., comps. 2004. Experimental forests and ranges of the USDA Forest Service. Gen. Tech. Rep. NE-321. Newtown Square, PA: U.S. Department of Agriculture, Forest Service, Northeastern Research Station. 178 p.

Bailey, R.G. 1991. Design of ecological networks for monitoring global change. Environmental Conservation. 18(2): 173–175.

Crawford, R.H. 2006. USDA Forest Service experimental forests and ranges. In: Irland, L.C.; Camp, A.E.; Brissette, J.C.; Donohew, Z.R., eds. Long-term silvicultural and ecological studies results for science and management. GISF Research Paper 005. New Haven, CT: Yale University: 222–225.

Kneipp, L.F. 2005. A national system of experimental forests and ranges. Science. 72: 560–561.

Lugo, A.E.; Swanson, F.J.; Gonzalez, O.R.; Adams, M.B.; Palik, B.; Thill, R.E.; Brockway, D.G.; Kern, K.; Woodsmith, R.; and Musselman, R. 2006. Long-term research at the USDA Forest Service's Experimental Forests and Ranges. Bioscience. 56: 39–48.

Miller, G.W.; Crawford, R.H. 2006. A national strategy for research and science delivery on experimental forests and ranges in the 21st century. Unpublished briefing paper.